Smart Public Safety Video Surveillance System

*To my beloved family, Jean, Soumady, Archana, Vijayalakshmi,
and Pecto, with all my love and gratitude.*
Abhishek Djeachandrane

To all my family, my wife, with all my love and gratitude.
Serge Delmas

*To my beloved wife, my wonderful children, and my dear family – your love
fills my life with joy and purpose. I am endlessly grateful for each of you.*
Said Hoceini

*To my grown-up daughter, Ikram, who has just married and whom
I wish full happiness in her new life. No matter how you
grow, you will always be my beloved little girl.*
Abdelhamid Mellouk

New Generation Networks Set
coordinated by
Abdelhamid Mellouk

Volume 5

Smart Public Safety Video Surveillance System

Innovative Technologies for Homeland Security and Mission-Critical Operations

Abhishek Djeachandrane
Said Hoceini
Serge Delmas
Abdelhamid Mellouk

ISTE

WILEY

First published 2025 in Great Britain and the United States by ISTE Ltd and John Wiley & Sons, Inc.

ISTE Ltd
27-37 St George's Road
London SW19 4EU
UK

www.iste.co.uk

John Wiley & Sons, Inc.
111 River Street
Hoboken, NJ 07030
USA

www.wiley.com

Library of Congress Control Number: 2025934394

British Library Cataloguing-in-Publication Data
A CIP record for this book is available from the British Library
ISBN 978-1-83669-054-2

Contents

Preface

In an era marked by rapid technological advancements and an increasing emphasis on public safety, the integration of intelligent systems into urban environments has become essential. The development of Smart Public Safety Video Surveillance Systems (SPSVSS) represents a significant step toward enhancing community security, empowering law enforcement officials to respond proactively to incidents while upholding civil liberties.

The journey presented in this book, the result of a collaboration between AIRBUS Defence and Space and UPEC (Paris XII University), stems from rigorous research and innovative engineering, focused on creating systems that intelligently analyze data in real time to monitor public spaces. Traditional surveillance technologies are insufficient for addressing the complexities of modern urban settings, highlighting the need for intelligent systems that leverage artificial intelligence, machine learning, and data analytics.

This work aims to bridge theoretical advancements with practical applications in public safety in order to develop advanced surveillance systems that adapt to their environments and provide critical insights for security personnel. Throughout this book, readers will encounter a comprehensive analysis of methodologies, challenges, and achievements.

A part of this this book was initially based on the work done in the framework of the Abhishek Djeachandrane PhD thesis, under the direction of Pr. Abdelhamid Mellouk. It was subsequently developed into a book to facilitate understanding of the technical aspects and the broader implications of surveillance technology. The findings originate from extensive lab work, collaborative discussions, experiments, and data analysis, helping readers understand key concepts and methodologies related to SPSVSS and stimulating dialogue about future advancements in the field.

To give a complete bibliography and a historical account of the research that led to the present form of the subject would be impossible. It is thus inevitable that some topics have been treated in less detail than others. The choices made reflect in part personal taste and expertise, and in part a preference for very promising research and recent developments in the field of technologies-based public safety, highlighting the ongoing need for research and development in this vital area.

This book is a start, but also leaves many questions unanswered. We hope that it will inspire a new generation of investigators and investigations.

The authors hope you will enjoy reading this book and get many helpful ideas and overviews for your own study.

Abhishek DJEACHANDRANE
Said HOCEINI
Serge DELMAS
Abdelhamid MELLOUK

April 2025

List of Acronyms

2D: Two-dimensional

3D: Three-dimensional

3GPP: The 3rd Generation Partnership Project

5G: Fifth Generation Technology Standard

6G: Sixth Generation Technology Standard

AC: Autonomic Computing

A-C: Actor-Critic

ACC: Accuracy

AdaBoost: Adaptive Boosting

ADC: Analog-to-Digital Converter

AI: Artificial Intelligence

API: Application Programming Interface

AUC: Area-Under-The-Curve

BCE: Binary Cross-Entropy

BMU: Best Matching Unit

C3D: Convolutional 3D/3D ConvNets

CCTV: Closed-circuit television

CE: Cross-Entropy

CLT: Central Limit Theorem

CNN: Convolutional Neural Network

CS: Camera Station

CVAE: Convolutional Variational Auto-encoder

DAC: Discriminant Analysis and Classification

DB: Database

DDoS: Distributed Denial-of-Service

DL: Deep Learning

DMIL: Deep Multi-Instance Learning

DRL: Deep Reinforcement Learning

DSS: Decision Support System

E2E: End-to-End

ECG: Electrocardiogram

EDA: Estimation of Distribution Algorithm

ELM: Extreme LearningMachine

EM: Expectation–Maximization

FAA: Federal Aviation Administration

FE: Feature Extractor

FFNN: Feed Forward Neural Network

FN: False Negative

FR: Full Reference

GB: Gradient Boosting

GCN: Graph Convolutional Network

GDPR: General Data Protection Regulation

GMM: Gaussian Mixture Model

GPS: Global Positioning System

GRU: Gated Recurrent Unit

HCAI: Human-Centered Artificial Intelligence

HCI: Human–Computer Interaction

HFST: High-Frequency and Spatio-Temporal

HMM: Hidden Markov Model

HOF: Histograms of Optical Flow

HOFO: Histograms of Optical Flow Orientation

HOG: Histograms of Oriented Gradient

I3D: Inflated 3D/Two-Stream Inflated 3D ConvNets

IFC: Industry Foundation Classes

IoT: Internet of Things

IP: Internet Protocol

IQA: Image Quality Assessment

IR: Information Retrieval

IT: Information Technology

IVS: Intelligent Video Surveillance

KL: Kullback–Leibler

KLT: Kanade–Lucas–Tomasi

KNN: K-Nearest Neighbors

LQ: Luminance Quality

LSB: Least Significant Bit

LST: Local Spatio-Temporal

LSTM: Long Short-Term Memory

MAB: Multi-Armed Bandit

mAP: mean Average Precision

MC: Monte Carlo

MCX/MCS: Mission Critical Communications

MDP: Markov Decision Process

MIL: Multi-Instance Learning

ML: Machine Learning

MLP: Multi-Layer Perceptron

MOS: Mean Opinion Score

MSE: Mean Square Error

MTL: Multi-Task Learning

MW: Mann–Whitney

NIQE: Natural Image Quality Evaluator

NN: Neural Network

NR: No-Reference

NYPD: New York Police Department

OWL: Web Ontology Language

PCA: Principal Component Analysis

PoC: Proof of Concept

PSNR: Peak Signal-to-Noise Ratio

QoE: Quality of Experience

QoS: Quality of Service

R-CNN: Region-Convolutional Neural Network

RDF: Resource Description Framework

RF: Random Forest

RNN: Recurrent Neural Network

ROC: Receiver Operating Characteristic

RoI: Region of Interest

RPCA: Region Principal Component Analysis

RTFM: Robust Temporal Feature Magnitude

SA: Situational Awareness

SL: Similarity Learning

SOM: Self-Organizing Map

SoTA: State of The Art

SS: Surveillance System

SSIM: Structural Similarity Index Measure

ST: Spatio-Temporal

STT: Spatio-Temporal Texture

SVM: Support Vector Machine

TD: Time-Difference

TF-IDF: Term Frequency-Inverse Document Frequency

TN: True Negative

TSN: Temporal Segment Network

tSNE: t-distributed Stochastic Neighbor Embedding

UAV: Unmanned Aerial Vehicle

VANET: Vehicular Ad-hoc Network

VCA: Video Content Analysis

vMOS: video Mean Opinion Score

VQA: Video Quality Assessment

WCAE: Weighted Convolutional Autoencoder

WCN: Wireless Camera Network

XGBoost: Extreme Gradient Boosting

Introduction

"The world is a dangerous place to live."
Albert Einstein

I.1. General context

In France, according to the French Ministry (2023), almost all recorded delinquency indicators were up in 2022 compared with the previous year. These increases follow those observed before the health crisis for homicides, intentional assault and battery, sexual violence and fraud recorded by the police and gendarmerie. For example, the number of recorded victims of intentional assault and battery (persons aged 15 or over) rose sharply in 2022 (+15%, after +12% in 2021).

Recorded crime indicators relating to non-violent theft from persons, burglary, vehicle theft, theft from vehicles and theft of vehicle accessories, which had fallen sharply during the health crisis, rose sharply in 2022.

According to statistics from the The French Central Directorate of the Judicial Police (2023), as shown in Figure I.1, the crime rate in France is on the rise compared to 2021. Against a backdrop of almost constant increases since 2008 (with a drop during the 2020 pandemic), however, with certain disparities, for example, vehicle theft is down as shown in Figure I.3, while physical and sexual violence are up as shown in Figure I.2.

For a color version of all the figures in this chapter, see www.iste.co.uk/djeachandrane/smart.zip.

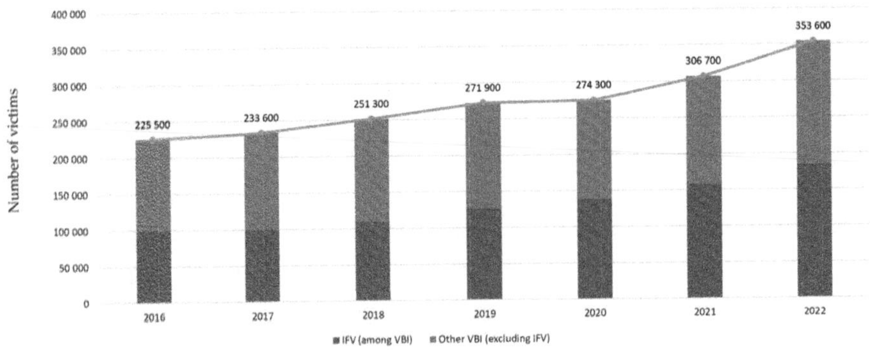

Figure I.1. *Voluntary blows and injuries among individuals aged 15 or older (VBI), and intra-family violence (IFV)*

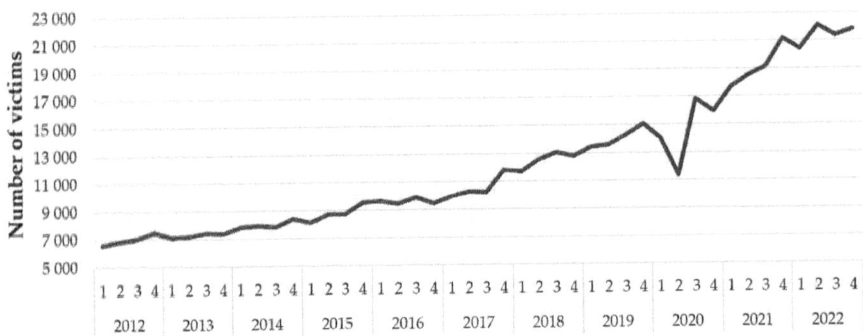

Figure I.2. *Recorded instances of sexual violence*

Statistics for specific types of violence, such as sexual violence or vehicle theft, attest to the existence of violence, even if trends are evolving differently, as shown in these respective figures, which show developments between 2008 and 2022.

This question of national security concerns not only countries such as France but also some important cities such as the world's largest city, New York, which is sounding the alarm.

For instance, according to the New York Police Department (2023), the apparent dips come after total major crimes spiked 22% last year, from 103,388 incidents in 2021 to 126,537 in 2022, as attested in Figure I.4 provided by the Office of the New York State Comptroller (2022) where the number of total crimes and total crime rates are rising for the first time in eight years.

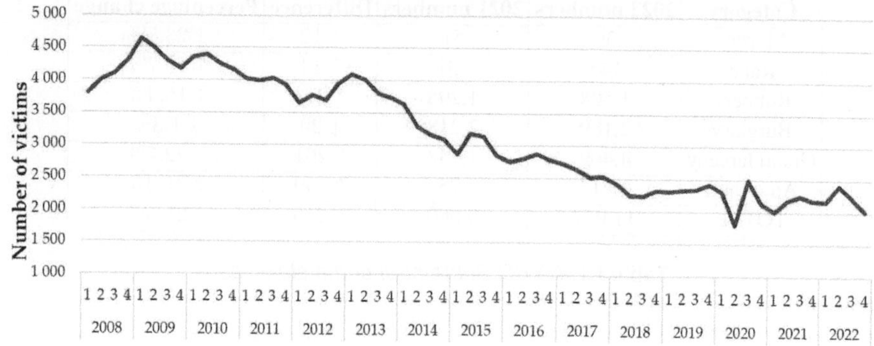

Figure I.3. *Recorded instances of vehicle theft*

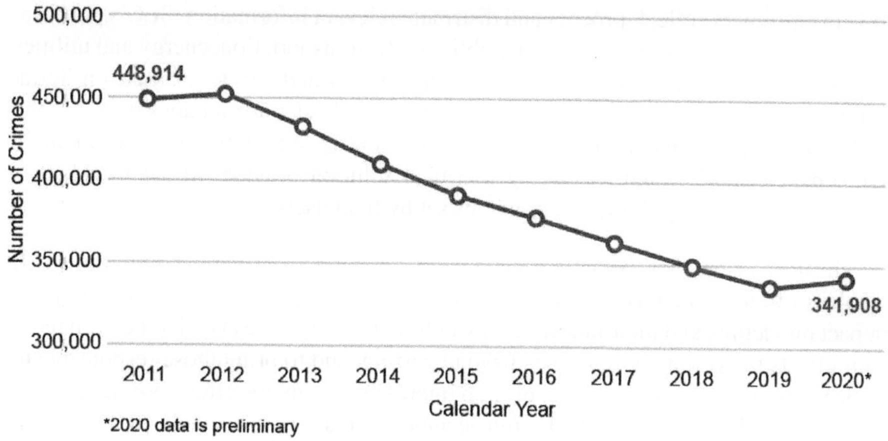

Figure I.4. *Total reported crimes in the state of New York*

As shown in Table I.1 by amNewYork Metro (2022), that rise was driven by increases in six of the seven categories: rapes, robberies, felony assaults, burglaries, grand larcenies and car thefts, according to NYPD statistics.

Criminal activity has been a persistent issue for societies throughout history. It takes on many forms and can occur in many different ways. It is important for law enforcement to be aware of these changes and to adapt their tactics to effectively combat crime and protect the public. Advances in technology have provided new tools for authorities to use in this fight against criminality, and it is important to make use of these tools to enhance public safety.

Category	2022 numbers	2021 numbers	Difference	Percentage change
Murder	39	51	↓ 12	↓ 23.5%
Rape	145	131	↑ 14	↑ 10.7%
Robbery	1,508	1,295	↑ 213	↑ 16.4%
Burglary	2,189	2,218	↓ 29	↓ 1.3%
Grand larceny	4,409	1,148	↑ 261	↑ 22.7%
Auto theft	1,215	3,753	↑ 799	↑ 21.3%
TOTAL	11,057	9,000	↑ 215	↑ 21.5%

Table I.1. *NYPD September crime statistics*

Indeed, the rising crime rate in France is a concern not only for law enforcement agencies but also for solution providers such asAirbus. Airbus is a global leader in aeronautics, space and associated services. One of its program units, secure land communications (SLC), offers advanced communication and collaboration solutions to help customers collect, process and distribute relevant information. SLC's portfolio caters to the needs of professionals in public safety, transportation, energy and utilities sectors. It includes networks, terminals, applications and services based on Tetra, Tetrapol and broadband technologies. As a European leader and a major player in the global market, SLC has customers in over 80 countries and employs approximately 1,150 people in 17 countries. Being a critical solution provider in the world, it is committed to meeting all the requirements set by final users.

Physical crime has been a long-standing concern for many communities. It involves offenses such as assault, robbery and homicide, and can have a significant impact on victims and their families. Law enforcement agencies need to be vigilant in their efforts to prevent these crimes from occurring, and to bring those responsible to justice. The use of technology can play an important role in this effort, providing tools such as surveillance cameras and forensic analysis to aid in investigations. Overall, it is crucial to take a proactive approach to combat physical crime and ensure the safety of the public.

Public safety refers to measures taken by governments and communities to protect citizens from harm and maintain order in society. This includes maintaining safe and secure communities, preventing and responding to crime and ensuring the safety of public spaces and infrastructure. Public safety is a critical component of smart cities as advanced technology and data analysis can be used to improve emergency response times, monitor crime trends and enhance overall safety for residents.

As mentioned, smart cities can potentially help to prevent crimes by implementing technologies such as surveillance cameras and predictive policing algorithms. These tools can aid in identifying and addressing criminal behavior in real time, thereby increasing public safety. Smart cities are urban areas that use advanced technology and data analysis to improve the quality of life for residents and enhance sustainability.

This includes using sensors, data analytics and automation to improve transportation, energy usage, waste management and public safety. Smart cities are becoming increasingly popular around the world as more and more people move to urban areas and governments seek innovative solutions to urban challenges.

For public safety purposes, video surveillance is a crucial aspect of maintaining security and safety in various settings. Whether it is in public places such as malls, airports or train stations, or in private locations such as homes or businesses, video surveillance systems can provide an extra layer of protection and deter potential criminal activity. Video surveillance can help prevent theft, vandalism and other criminal acts. It can also aid in identifying suspects and providing evidence for investigations. In addition, video surveillance can help monitor employee behavior, improve workplace safety and ensure compliance with regulations and policies. Moreover, video surveillance can also be used for remote monitoring, allowing for real-time observation of events or incidents that require immediate attention. With advancements in technology, video surveillance systems can now incorporate features such as facial recognition, license plate recognition and motion detection, enhancing their effectiveness and accuracy.

Overall, video surveillance plays a vital role in maintaining safety and security, providing valuable insights and enabling quick responses to potential threats or incidents. For all the reasons mentioned, video surveillance is considered a critical service because it plays a crucial role in ensuring the safety and security of people and property. It can be used to monitor and detect potential threats, prevent crime and provide evidence in the event of an incident. In many cases, video surveillance systems are integrated with other security measures to create a comprehensive security solution. The reliability and effectiveness of these systems is paramount, as they can have a significant impact on public safety and the protection of valuable assets.

A general description of video surveillance will allow us to better identify principal challenges, by giving a general overview of video surveillance systems. In this area, the end devices in the architecture are considered the eye of the system; the communication media are designed as synapses to convey information and the analytics represent the brain of the system. Here, video surveillance will be defined by architecture and analysis, the former consisting of network architecture and computing infrastructure, the latter of all the analysis that can be produced from the information gathered.

In video surveillance architecture, network architecture could be defined as follows.

– Analog surveillance systems: analog signal processing technology is the basis model for image transmission, exchange and recording using a short-distance coaxial cable and a long-distance transceiver optical fiber.

– Digital surveillance systems: by introducing digital video recorders (DVR), digital image files are transmitted. This differs from analog video surveillance in these aspects: signal transmission, control and storage. Video encoding formats like MPEG-4 and H264 enable digital video surveillance systems to conduct image monitoring with remote transmission across all existing digital networks while utilizing low bandwidth.

– Network surveillance systems: the network system is based on digital signal processing and involves network cameras or IP cameras, which is a type of digital video camera. The networking technique is used to realize signal transmission, exchange, control and video storage. All the components could be integrated into the system in a wireless or wired way to tackle many problems on this topic.

The key difference between these architectures is in transmission, exchange and storage. The main difference between digital and network surveillance systems is that digital surveillance systems (SS) unlock new capabilities of digital signal processing contrary to analogical signal, and network SS unlock new capabilities of processing through Internet protocol. In general, digital SS convert analogical signals to digital ones, while the network ones deal only with digital signals.

Computing infrastructure is where data from edge devices is processed and stored and could be defined as:

– Cloud computing: cloud computing refers to the practice of accessing computing resources, such as servers, storage and applications, over the Internet instead of owning and maintaining them locally.

– Fog computing: refers to a decentralized computing architecture that involves using localized devices to bring computing resources closer to the edge of the network, which can help reduce latency, improve security and lower bandwidth usage.

– Edge computing: edge computing is a distributed computing paradigm that involves processing and storing data at or near the edge of the network, where the data is generated. It typically involves using more powerful devices.

– Local computing: local computing usually refers to computing at one premise and is organized as a local area network.

The key difference between these infrastructures is in the localization of intelligence and computing power.

In video surveillance analytics, these tasks and goals could be defined as:

– face detection or recognition;

– human action detection or recognition;

– object detection and tracking;

– crowd counting;

– video summarization;

– abnormal event detection;

– traffic road congestion/accident detection;

– car plate recognition;

– abnormal behavior detection.

One of the primary responsibilities of an automated surveillance system is to monitor the environment being surveilled through the use of cameras, identify and evaluate the observed situation efficiently, and subsequently alert appropriate personnel if necessary. All of this is done by its analytics components which could be modeled by:

– traditional computer programs;

– computer programs coupled to image processing techniques;

– Artificial Intelligence (rule-based or data-driven) and its computer vision domain.

The key difference between these analytics solutions is in their information source and goal complexity.

On the one hand, the implementation of smart technologies such as surveillance cameras and predictive policing algorithms can potentially help reduce crime rates and increase public safety. On the other hand, concerns have been raised about the potential for these technologies to infringe on privacy and civil liberties governed by the GDPR in Europe, for example, as well as the potential for bias and discrimination in the use of predictive policing algorithms.

I.2. Challenges

With the evolution of technologies (exponential increase in data exchange capacities with 4G and now 5G, with 6G announced for the end of the 2020s, or the advent of the IoT phenomenon, which is still in its infancy), the multiplication of information sources and flows (open sources, social networks, IoT sources, etc.), and the availability of increasingly powerful and diverse terminals and accessories (smartphones, tablets, connected watches, cameras, augmented and virtual reality devices (e.g. glasses), biometric and environmental sensors, etc.), administrations, organizations and security agents (public security, civil security, private security) in the field and in command centers have, and will increasingly have, access to a considerable amount of information whose "intelligent" exploitation could enable disruptive advances in the accomplishment and effectiveness of their missions.

What is more, we need to bear in mind that the agents concerned may be operating in a wide variety of conditions (pedestrian or vehicular agents, or in a command center or crisis call management), in difficult or even hostile environments and in stressful situations (hands busy, driving, attention required on a critical task or potential danger).

The interaction of these professional users with real-time information flows in these constrained operational conditions therefore requires the use of innovative technologies such as Artificial Intelligence, providing them with, among other things, mission-specific information analysis and presentation and decision support capabilities, as well as operational efficiency.

I.3. Objectives

The field of application of Artificial Intelligence technologies to security missions, as outlined in the previous section, is vast. The work targeted by the book will be essentially related to the four themes developed below.

Augmented situational information. Situational information is to be understood here in the sense of situational awareness, i.e. the perception and understanding of the spatio-temporal environment and the events taking place within it, and their possible projection into the future. It is an essential element in decision-making in a number of fields, notably for control, safety and rescue missions, whether in the public domain (public safety, civil security) or the private sector (transport, industry, essential services). As mentioned in section I.2, the evolution of technologies and information sources means that security and rescue professionals now have access to vast quantities of multimedia (data, voice, video), multi-source (IoT sensors, cameras, command centers, specialized applications, public networks, open sources, social networks, the web, the dark web, etc.), real-time and non-real-time data. The main objective is therefore to be able to use this data to build up "augmented" real-time (or near-real-time) situational information using Artificial Intelligence technologies in a way that is relevant and contextual to the mission and situation, integrated with mission control and execution tools.

Operational decision support. Complementing (or coupled with) "augmented situational information" (see the relevant paragraph above), based on the same data and information, possibly supplemented by other sources of intelligence, drawing on business procedures and their historical data (often kept for legal reasons), real-time (or near-real-time) decision support functionalities using Artificial Intelligence technologies, relevant and contextual to the mission and situation, integrated with mission control and execution tools, would be a major added value for both command room personnel and agents deployed in the field.

Intelligent interaction. Users can operate in a wide variety of conditions (pedestrian or vehicular agents, or in a command center or crisis call management), in difficult or even hostile environments, and in stressful situations (hands busy, driving, the attention required on a critical task or potential danger). The equipment and means of interaction with a growing mass of information, sophisticated devices and applications developed for the general public are, therefore, generally not adapted to the operational uses of safety and rescue professionals. The main objective is therefore, thanks to Artificial Intelligence technologies, possibly combined with other technologies (e.g. augmented reality or voice recognition), to be able to offer the users concerned the possibility of benefiting from new capabilities by simplifying interactions with devices and equipment (e.g. augmented reality display, voice recognition/voice command in difficult environments and stress conditions, etc.) in a way that is adapted to and integrated with their operational environment (equipment and applications of control room agents and first responders in the field).

Assisted procedure management and reporting. Safety and rescue personnel devote a significant proportion of their time to low value-added tasks (procedural, administrative, repetitive time-consuming tasks). Artificial Intelligence technologies would appear to be a well-suited means of providing effective solutions in this field, for example, for intelligent assistance in the management of "workflows" or intervention reports, integrated with mission control and execution tools, enabling users concerned to be more efficient and have more time available for their missions, while alleviating the tedious and low-value aspects of their job and minimizing the risk of error.

I.4. Problematics and contributions

In critical operations, the complexity of surveillance tasks can often overwhelm the dedicated public safety personnel. Fortunately, today's arsenal of intelligent tools can help decision-making. In the state of the art, cutting-edge technologies have the potential to gather the necessary context from vast volumes of data and extract essential insights for arming decision-makers with invaluable information. These decision-makers anticipate that intelligent video surveillance systems will deliver a level of reliability comparable to human judgment. Achieving this hinges on the system performance aligning with user expectations for all implicit decision-support tasks. In addition to their core responsibilities, decision-makers are often burdened with low-added-value tasks such as system management. Regrettably, these secondary background tasks are time-sensitive, involving critical actions such as remediation or configuration tasks, and any delay in their execution can yield catastrophic consequences.

Based on this reasoning, the following three questions arise: **How to assess and provide a control signal to the system? How to draw conclusions from the surrounding context? And how to improve the responsiveness of system management?**

In response to these inquiries, a comprehensive framework has been meticulously designed and proposed as a development platform. The framework encompasses three fundamental strategies.

1) Firstly, it introduces an end-to-end corrective signal that considers exogenous, endogenous and human factors of the surrounding context. This signal is inspired by the data enrichment process. To assess the overall system, a task-specific Quality of Experience (QoE) measure is provided.

2) Secondly, the solution advocates for the implementation of predictive systems rooted in machine learning and inspired by situational awareness. These systems aim to enhance system capabilities in the first instance and result-oriented performance in the second. This approach addresses real-world scenario problems and allows for the temporal localization of anomalies, adaptation to concept drift, learning from no example and model interpretation.

3) Lastly, the solution suggests the adoption of zero-touch self-management systems, taking inspiration from the principles of autonomic computing. These systems are designed for configuration, protection and learning, and are based on machine learning, descriptive and inferential statistics, and control theory. These solutions reduce user requests, increase system responsiveness to management and improve system acceptability.

Collectively, these approaches provide a robust and comprehensive framework for addressing a range of critical questions by leveraging a variety of cutting-edge technologies and methodologies. By combining data enrichment, situational awareness and autonomic computing, this framework is well suited to meet the demands of modern businesses, where the ability to learn, infer and adapt quickly is more important than ever.

I.5. Structure of the book

This book is organized as follows:

Chapter 1 provides a new literature review on an "End-to-End Video Surveillance System for Public Safety". The purpose is to review analytics methods for human threats in the cyber and physical space to better identify drawbacks of existing methods.

Chapter 2 presents the development platform for integration and testing. The purpose is to provide a high-level view of the system and a detailed framework and to define the Airbus DS SLC use case to unlock a new perspective.

Chapter 3 describes a novel multi-criteria enriched corrective signal. The purpose is to provide information about surrounding awareness in real-world scenarios regarding endogenous, exogenous and human factors.

Chapter 4 describes a novel situational awareness-centric predictive system. The purpose is to augment this system's capabilities to meet mission-critical requirements and to reach satisfying results for anomaly detection to operate in real-world situations.

Chapter 5 describes a novel autonomic approach for intelligent video surveillance systems. The purpose is to let the system self-manage to gain in ownership and performance by reducing user managing time.

The final chapter concludes this book and provides an insight into our future work.

1

Literature Review on an End-to-End Video Surveillance System for Public Safety

"Technology is a useful servant but a dangerous master."
Christian Lous Lange

Chapter 1 presents a new literature review on an "End-to-End Video Surveillance System for Public Safety". The purpose is to review analytics methods to human threats in the cyber and physical space to better identify drawbacks of existing methods.

Concerns about human threats in urban areas are increasing, in parallel with the dramatic growth in rates of violence, incivility and crime. At the same time, with the recent breakthroughs in analytic approaches and video surveillance architecture, CCTV can gather and process a wealth of sensitive information from different sensors for a more valuable analysis task. Combating human threats in urban, cyber and physical space is an important and challenging topic. However, to date, no systematic studies have been carried out on the subject of an end-to-end video surveillance system for public safety. In this literature review, we first discuss threats in cyber and physical space, so that the analysis can identify threats in physical states, and, following the architecture, we discuss how the information will be transmitted to the analysis component. Next, we look at the impact that cyberspace – the architecture – and physical space – the analysis – could have. Given the critical nature of video surveillance applications, it is imperative that all studies take into account the perspective of the end-user, who is the ultimate beneficiary of this service. Therefore, the focus of all research should be on the user experience and their needs and expectations.

For a color version of all the figures in this chapter, see www.iste.co.uk/djeachandrane/smart.zip.

1.1. General description: human threats in urban areas and abnormal situation detection

Human threats in urban areas can come in various forms. One of the most common threats is crime, which includes theft, burglary, assault and other forms of violence. Another threat is pollution, which can harm people's health and the environment. There is also the danger of accidents, such as car crashes or building collapses. Additionally, overcrowding can lead to social and economic problems, including poverty, unemployment and homelessness. These issues can have a significant impact on the quality of life in urban areas, and it is important for policymakers and communities to work together to address them. These human threats could be identified as the most common threats in physical space. Initially, threats in physical space, particularly in urban areas, came from asymmetric military threats (Pavlidis et al. 2001). The next challenge is to think about future threats whose impact could be similar to that of physical threats, i.e. cyber threats.

Figure 1.1. *Urban area-specific direct human threats*

1.2. Analytics for video surveillance

Analytics could be done by well-known image processing methods or computer vision methods. Image processing is mainly focused on processing the raw input images to enhance them in order to do other tasks, while computer vision is focused on extracting information from the input images or videos to have an understanding of them in order to predict the visual input such as a human brain. Image processing tasks relate to rescaling, correcting illumination change tones, etc., while computer vision tasks relate to object detection, face detection, handwriting recognition, etc. Image processing is a subfield of computer vision. Advanced analytics require computer vision techniques instead of only traditional image processing methods. Computer vision could be supported by computer programming or machine learning models, also called data-based AI. Computer programming relates to simply coding a conditional code by fixing a threshold, for example, or detecting a class by comparing distance and by fixing a threshold. And machine learning will consist of learning certain behavior without rules provided by a human. In traditional programming, the program uses data in a logical way to achieve a task, while machine learning uses data to arrive at a logic that can further be used to predict patterns. Some advanced computer programming methods could be referred to as rule-based AI with more complex components, such as inference engines or knowledge representation methods.

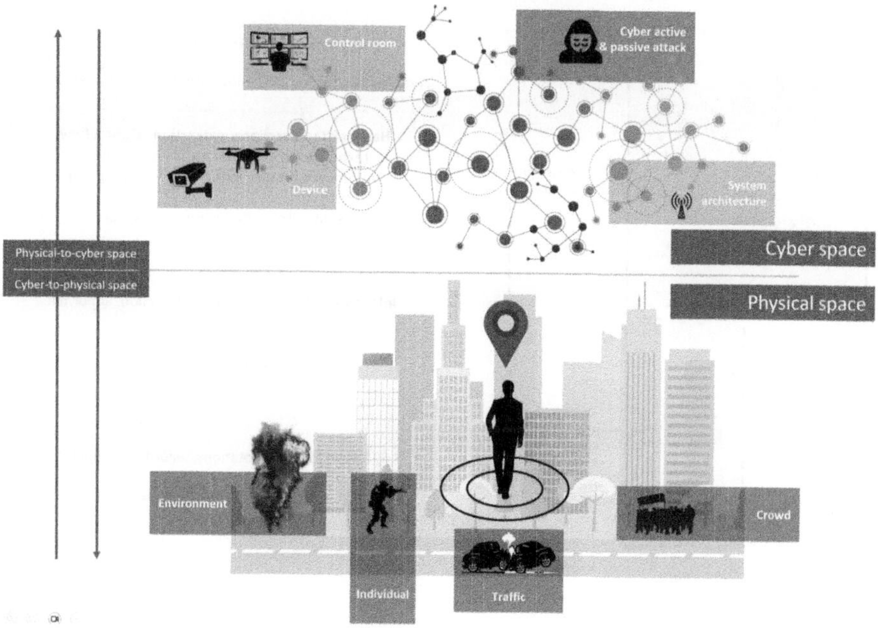

Figure 1.2. *Human threats in urban areas from cyber and physical space*

In data-driven AI, simple machine learning and deep machine learning exist, also defined by their model: shallow or deep. Using a shallow model consists of providing handcrafted features, and the shallow model will learn to produce an expected result, while deep learning does not even need handcrafted features; it could learn it automatically and enable it to learn from unstructured data directly without any intermediary transformation.

Many combinations could exist by using image processing or computer vision methods in addition to traditional programming and/or AI methods.

For our literature review, an analysis framework is proposed in Figure 1.3 to present the important points; all methods will be analyzed through this framework, and the granularity of the analysis will depend on the necessary discrimination criteria of comparison. The data pipeline will highlight all the methods used to describe the data, while the knowledge modeling will show how the knowledge has been modeled, containing the model structure and learning schema, as well as prior knowledge about the data in the form of labels. Finally, reasoning will highlight the inference mechanism possible thanks to the combination of data and knowledge.

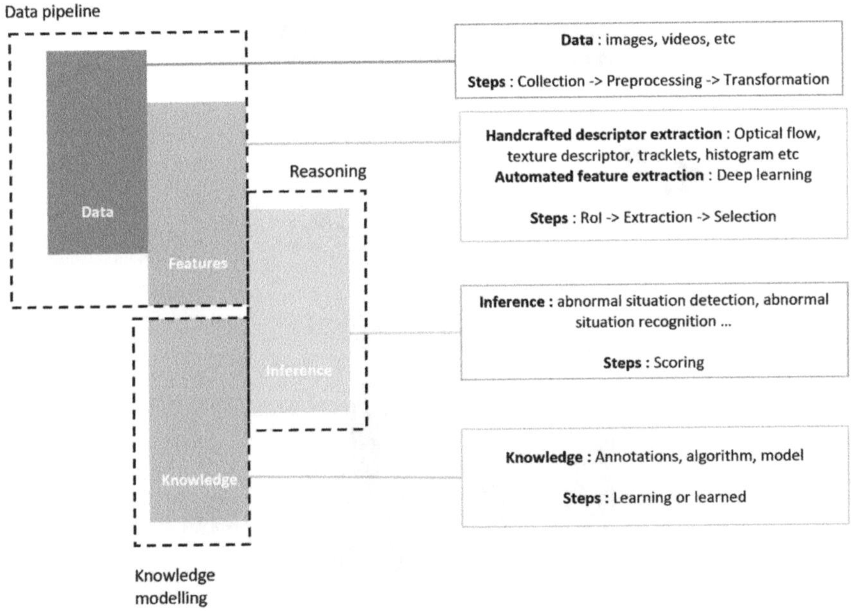

Figure 1.3. *Data-driven method framework*

The literature review will be carried out on analytics in human threats in urban areas; this has been done by studying analytics in abnormal behavior analysis in the crowd, traffic, environment and individuals.

1.2.1. *Crowd behavior analysis*

As presented in this survey (Li et al. 2015), the analysis of the crowd can be conducted at macroscopic or microscopic levels. At the macroscopic level, the global motions of a mass of people are focused, without concerning the movements of any individual; at the microscopic level, the movements of each individual pedestrian are concerned and analyzed based on their collective information. For the crowd scene anomaly detection, feature representation and prediction models should be set to these applications. Feature engineering is important in image-related analysis tasks. Feature extraction could be handcrafted or based on deep learning. Then, classification could be done by shallow or deep models, too. Distinguishing these two phases is crucial to understand the crowd dynamic and how to discriminate it. First, motion representation and event detection will be covered in the same order as the crowd scene.

Motion representation

As a crucial basis, an appropriate feature representation can benefit the subsequent tasks. Motion information representation is the basis for crowd scene analysis. Although other types of visual features, such as scene structure, geometrical information and viewing direction, could also be helpful, motion features are dominant in our area.

Handcrafted features

According to the representation level, previous crowd motion features can be categorized into three groups: flow-based features, local spatio-temporal features and trajectory/tracklet. Flow-based features are extracted densely on the pixel level. Local spatio-temporal features represent the scene based on local information from 2D patches or 3D cubes. On a higher level, trajectory/tracklet tries to compute the individual tracks as the basic features for motion representation. These feature representations have been used in various tasks such as motion pattern segmentation, crowd behavior recognition and crowd anomaly detection. These features could be classified into two modeling classes: macroscopic and microscopic. Optical flow and spatio-temporal features have been classified in the macroscopic modeling method, and tracklet features are classified in the microscopic modeling method.

Deep learning-based feature extraction

This method is an automated way to extract features according to a specific task. The process of extracting features of data from the deep layers of a neural network is referred to as deep feature extraction, and the features extracted are called deep features. These features could be extracted through all existing neural network architecture. From traditional Feed Forward Neural Networks (FFNN) to encoder–decoder approaches, through Convolutional Neural Networks (CNN) for spatial representation as images, and Recurrent Neural Networks (RNN) for time series as ECG signals or both CNN–RNN for temporal, spatial representation as videos, even using attention mechanism-based neural networks.

Ref.	Feature Extractor			Classifier		
	RoI	**Extrac.**	**Select.**	**Model**	**Learn.**	**Result**
Wang et al. (2012)	Atoms in general	High-frequency spatio-temporal features: Local ST cuboid and wavelet transform	PCA	HMM	offl.	90% AUC (UMN) (survey + own dataset)
Wang et al. (2018b)	Atoms in general	Optical flow	Histogram of optical flow orientation (HOFO)	HMM	offl.	96.17% AUC (UMN)
Feng et al. (2010)	Atoms in general	Optical flow	Gaussian distribution over normal behavior flow vectors	SOM with KL divergence distance	offl.	87.8% AUC (UMN)
Hao et al. (2018)	Density estimation with Gabor filter on STT	Atoms in general	Contrast, angular second moment, entropy, variance	KNN, Naive Bayes, DAC, Random Forest, SVM	offl.	63-87% ACC (own dataset)
Gao et al. (2018)	Atoms in general	Optical flow	Persistence, direction, motion magnitude	Weighted Random Forest	offl.	95% AUC (UCSD Ped1)
Nain and Lamba (2018)	Chan-Vese Segmentation	Tracklets with KLT tracking algo	Shape, location, flow direction	KNN	offl.	99.5% AUC (UMN) 85% AUC (UCF Web) 95% AUC (Violent Flows)

Table 1.1. *Comparative study of handcrafted methods in crowd scene analysis*

Event detection

Once these features are either handcrafted or automatically extracted from them, classification could be done with shallow or deep models. If you can move mountains, you can move molehills, but not the inverse. Because, from handcrafted features, low-dimensional data representation is obtained due to statistical information of handcrafted methods, but from deep learning, high-dimensional data is obtained to preserve information for the discrimination, so the classification could be tougher. When low-dimensional, the shallow model is employed, and when high-dimensional, the deep model is used. A deeper analysis of some existing methods is detailed in Table 1.1 for handcrafted methods and in Table 1.2 for deep learning methods.

Ref.	**Feature Extractor**			**Classifier**		
	RoI	**Extrac.**	**Select.**	**Model**	**Learn.**	**Result**
Xu et al. (2019)	CVAE and multiple Mixture Gaussian Model				offl.	95.7% AUC (UCSD ped1) 92.3% AUC (UCSD ped2) 86% AUC (Shanghai tech)
Yang et al. (2018a)	Moving foreground with RPCA + WCAE-LSTM				offl.	85% AUC (CUHK Avenue) 85.1% (UCSD ped2) 92.6% AUC (UCSD ped1)
Singh et al. (2020)	Transfer Learning in CNNs network (AlexNet, GoogLeNet, VGGNet) and Aggregation of Ensembles of fine-tuned ConvNets + Softmax and SVMs classifier				offl.	94.6% AUC (UCSD ped1) 95.9% AUC (UCSD ped2) 89.3% AUC (CUHK Avenue)

Table 1.2. *Comparative study of deep learning methods in crowd scene analysis*

In handcrafted feature extraction, using listed methods is not efficient; defining a region of interest (RoI) and selecting the relevant features that describe the event is important. In Wang et al. (2012), high-frequency and spatio-temporal (HFST) information computed by the wavelet transformation to characterize the dynamic properties of the local region are used for feature extraction and, in addition to this, multiple Hidden Markov Models (HMM) are used to detect various local abnormal crowd events. Multiple HMMs are used to model behaviors, one HMM by behavior. In Wang et al. (2018b), they derived histograms of optical flows (HOF) from optical flow feature extraction to model in an unsupervised way normal frames through an HMM. Then, they measure the similarity between normal and abnormal frames. In Feng et al. (2010), they use optical flow features, then model optical flow vectors of normal behavior into Gaussian distribution to smooth outliers and then use an unsupervised method called a self-organizing map (SOM) to represent the behavior. A simple similarity function has been defined in order to classify. The Kullback–Leibler (KL) divergence is used as a distance metric to calculate the similarity between normal and abnormal behaviors where the threshold has to be fixed manually. In Hao et al. (2018), more than its spatial–temporal texture analysis, the Gabor filter allows for getting any specific frequency content in the image in specific directions in a localized region around the point or region of analysis and then applying gray level co-occurrence matrix (GLCM) on spatio-temporal texture (STT) permits us to characterize the texture of an image by calculating how often pairs of pixels with specific values and in a specified spatial relationship occur in an image, and then extracting statistical measures from this matrix. From provided statistical information, the authors classify them with shallow ML models as K-Nearest Neighbors (KNN), Naïve Bayes, Discriminant Analysis Classifier (DAC), Random Forest (RF) and

Support Vector Machine (SVM), which is possible due to their low-dimensional feature representation, also called describers. In Gao et al. (2018), using optical flow is not enough; here, RoI are not considered, they take all information as atoms and, from the optical flow, use some statistics extractable from it as persistence, direction or motion magnitude. Then, classify them with a weighted random forest and obtain a result of 95% AUC on UCSD peds 1. In Nain and Lamba (2018), using tracklets is not sufficient; the RoI is obtained by Chan–Vese segmentation methods, and then from tracklets providing information, shape, location and flow direction, they classify them with KNN, which obtained a result of 99% AUC, on UMN and 85% on UCF Web. For a deep learning-based approach, the RoI does not have to be set; it will be automatically defined and features will be extracted from it automatically. More context is provided, and it will be accurate instead of limiting it by the selected RoI-handcrafted way. Here, feature extraction and classification are processed in the same model. The method (Xu et al. 2019) uses a Conv-Variational AutoEncoder (CVAE) which consists of obtaining a latent space that contains extracted features and then are classified in an unsupervised way, i.e. when it does not belong to this model, a low probability will show it as abnormal, so a threshold should be set, and obtained a result of 95% on UCSD. In Yang et al. (2018a), CNN-based AE in an RNN is used to learn spatial representation in a sequential way and then use a regularity score to reflect the probabilities of abnormal events inspired by peak signal-to-noise ratio (PSNR) from reconstructed data and ground truth and obtain a result of 95% AUC on UCSD. In Singh et al. (2020), they use a multitude of SoTA CNN models using transfer learning on AlexNet, GoogLeNet and VGGNet and proceed to aggregation of ensembles with a softmax classifier and linear, quadratic and cubic SVM and provide a result of 95% AUC on UCSD.

Approach	Learning schema	Learning model	Ref.	Year
Handcrafted	Supervised	Unsupervised	Wang et al. (2012)	2011
	Supervised	Unsupervised	Wang et al. (2018b)	2018
	Unsupervised	Unsupervised	Feng et al. (2010)	2010
	-	Supervised	Hao et al. (2018)	2019
	-	Supervised	Gao et al. (2018)	2018
	-	Supervised	Nain and Lamba (2018)	2018
Deep learning	Unsupervised	Unsupervised	Xu et al. (2019)	2019
	Unsupervised	Unsupervised	Yang et al. (2018a)	2018
	-	Supervised	Singh et al. (2020)	2020

Table 1.3. *Comparative study of learning scheme and model in crowd scene analysis*

In Table 1.3, methods that use handcrafted feature principles were assessed with the UMN dataset (Mehran et al. 2009), which is the less challenging dataset among all of those listed. Although it was used in recent years, as in Wang et al. (2018b), it was assessed with UMN because we suppose that that approach is not competitive with others, i.e. with a new dataset introduced with more complexity, where complexity refers to resemblance to real scenarios. Most of the deep learning approach was used on datasets with more challenges, such as UCSD (Mahadevan et al. 2010) and CHUK (Shao et al. 2014). Rarely are handcrafted approaches able to compete with deep learning approaches, as with the method by Gao et al. (2018) done on UCSD ped1. In handcrafted methods, slightly unsupervised methods perform less well than supervised methods. Among unsupervised models, the learning schemes initially unsupervised could also be supervised, as done in Wang et al. (2012, 2018b); clearly, the methods that use a clear unsupervised learning scheme and models are able to complete a supervised learning scheme with unsupervised models or supervised models. Another research direction is defined in some of the studies using deep learning. They exploit deep learning models by assessing their models' ability to generalize over a dataset and to perform to another one, which clearly shows their limit in handcrafted methods by omitting this kind of study. In deep learning approaches at this stage, unsupervised or supervised methods perform well in modeling crowd behavior.

Approach	Modeling type	Method	Ref.
Handcrafted	Macroscopic	LST	Wang et al. (2012)
	Macroscopic	Optical flow	Wang et al. (2018b)
	Macroscopic	Optical flow	Feng et al. (2010)
	Macroscopic	LST	Hao et al. (2018)
	Macroscopic	Optical flow	Gao et al. (2018)
	Microscopic	Tracklets	Nain and Lamba (2018)
Deep learning	Hybrid	NN	Xu et al. (2019)
	Hybrid	NN	Yang et al. (2018a)
	Hybrid	NN	Singh et al. (2020)

Table 1.4. *Comparative study of feature modeling in crowd scene analysis*

In Table 1.4, we show the feature modeling level – macroscopic or microscopic – to show that the majority of methods in our review are macroscopic in handcrafted methods and that, in the deep learning approach, it was generally the hybrid approach; the focus on macro or micro will depend strongly on the optimization function. Some studies can strengthen more macroscopic or microscopic feature analysis in deep learning, but the final goal is to automatically extract relevant features for a specific task. This is the latter which gives the possibility of generalizing the deep learning model to other datasets.

1.2.2. *Traffic analysis*

Traffic anomalies are also an important category of abnormal urban events and are strongly linked to public safety. Urban analysis can come from a variety of sources; we are interested in abnormal behavior related to human threats, which can be due to malicious or unintentional actions. For example, there are many types of threats, such as road accidents involving a vehicle, a person, a bicycle or even a run-off.

Ref.	Approach	Method	Result
Robles-Serrano et al. (2021)	Deep learning	CNN InceptionV4 + ConvLSTM	98% ACC (CADP + UA-DETRAC dataset)
Zhou et al. (2022)	Deep learning + shallow model	Contain a temporal encoding and clustering to localize temporally the anomaly and proceed to spatial encoding and classification. - Temporal encoding: HOF + RGB automated feature extraction by MLP - Clustering: Self-Representation Constrained Low-Rank Representation - Spatial encoding: R-CNN - Classification: SVM	74.34% Mean AUC (DoTA public dataset)
Pramanik et al. (2021)	Image processing and shallow machine learning	Object detection: spatial-temporal neighborhood granulation/clustering. Object tracking: tracklet by Kalman filter. Speed Violation Detection: Dynamic feature as velocity from tracklet velocity. Overtaking detection: Extract the direction of velocity and define the RoI of the road. Illegal parking detection: HOG and SVM. Wrong drop-off location of passengers: SVM and HoG.	89.4% mACC (CamSeq01 dataset and Illegally Parked Vehicle Dataset (ISLab-PVD) and traffic videos acquired from a plant in India)
Aboah et al. (2021)	Deep learning and shallow model	Yolov5 for foreground and background detection and decision tree	F1 score of 0.8571 (Data from NVIDIA AI CITY CHALLENGE 2021 Track 4)

Table 1.5. *Comparative study of methods in traffic analysis*

In Table 1.5, the method presented by Robles-Serrano et al. (2021) consists of developing a completely automated method based on deep learning to detect car accidents. The method presented by Zhou et al. (2022) consists of using the computer vision method and the deep learning approach to proceed to spatial and temporal features encoding with MLP and doing classification with a shallow model called SVM to classify more complex car accidents. The method presented by Pramanik et al. (2021) consists of developing a set of methods based on simple image processing, computer vision methods at first as object recognition, tracklets extraction and histogram of oriented gradient (HOG) and doing classification with a shallow model called SVM of the pre-accident event as speed violation, overtaking and others. The method presented by Aboah et al. (2021) consists of developing a method based on deep learning for feature extraction and a shallow model called a "decision tree" to classify car stoppage.

There is currently no agreement on the best methods or approaches to use for data exploration due to a lack of available data. As a result, various processing combinations have been developed to accomplish similar tasks. Additionally, there is no consensus on the goals to achieve or how to describe the context. In order to collectively take advantage of a dataset and provide a benchmark for proposed methods to advance research, it is essential to define the dataset.

1.2.3. *Environment analysis*

Environment anomalies are also a significant category of anomalous urban events and are strongly related to public safety. Urban analysis can come from a variety of sources; we are interested in abnormal behavior related to human threats, which can be due to bad intentions or unexpected natural phenomena. For example, there are many types of threat, such as arson, natural fires and various types of explosions that could occur at gas stations, flooding caused by rain or river overflow.

The method presented by Munawar et al. (2021) (Table 1.6) consists of developing a deep learning model based on CNN to extract features and classify them from aerial data in the 2D plane to detect river flood, also called river overflow. The method presented by Ruiz and San Miguel (2016) uses traditional image processing and computer programming methods to detect rainfall-based floods from CCTV cameras by performing a binary thresholding of the difference for blob detection purposes. The method presented by Kim and Lee (2019) permits the detection of flames or smoke from a deep learning-based object recognition method and a sequence learning, where spatial and temporal knowledge is fused at different levels of the model to mix short- and long-term contexts. The method presented by Qiao et al. (2020) rethinks traditional image processing by using a deep learning method to extract the optical flow and proceed to adaptive thresholding segmentation to detect natural disasters, tsunamis and landslides from satellite views. The method presented by Lin et al. (2019) proposed a Faster R-CNN mixed 3D CNN to extract features and SVM to discriminate anomalies as smoke.

For all these methods, the dataset was composed by the authors themselves, and no reference dataset was provided to compare their method to existing ones. Moreover, this was due to the sparse anomalies that the authors are looking to detect from fires to tsunamis and landslides by passing through floods. Since 2016, traditional image processing methods have evolved into deep learning methods and some methods, such as the one presented by Qiao et al. (2020), developed deep learning methods to reproduce image processing analysis instead of using deep learning as in others' approaches to doing end-to-end global analysis until detecting the anomaly.

Ref.	Approach	Method	Result
Munawar et al. (2021)	Deep learning	CNN	91% (own dataset) 2150 images on a flood-prone region of the indus river in Pakistan.
Ruiz and San Miguel (2016)	Image processing and computer programming	Program which discriminates according to blob size to area of interest, compare histogram of current frame and referenced frame and mean pixel intensity of current frame and referenced frame. The area of interest and the referenced image were manually provided.	Tested on four video sequences and was able to react in a timely manner appropriate for its situation
Kim and Lee (2019)	Deep learning	Faster R-CNN and LSTM and spatial and temporal knowledge fusion for decision	97.92% mAP (own dataset) Collected 73,887 still images for CNN and collected 1,309 video clips from YouTube.
Qiao et al. (2020)	Deep learning based traditional image processing	Optical flow estimated by CNN and changed area segmentation based on an adaptive threshold selection	Greater than 95% ACC for both datasets (Satellite view dataset 301 video clips of tsunami in Petobo, Indonesia and 172 video clips of slow-moving landslide)
Lin et al. (2019)	Deep learning	Faster R-CNN + 3D CNN and SVM	95.23% ACC (own dataset smoke detection) 58 videos

Table 1.6. *Comparative study of methods in environment scene analysis*

1.2.4. *Individual behavior analysis*

The study of crowd behaviors and detection of violent actions in videos are well-established research fields, with significant advancements in violence detection since the first paper using a deep learning approach appeared in 2014. Some datasets even exhibit 100% accuracy in violence detection for short clips with labels for the entire clip. To push the limits further, researchers have created new testbeds with hours of CCTV camera recordings, longer videos and frame-level annotations. Additionally, some of these testbeds have moved from binary classification to multiple classes, differentiating the typology of violence in the video. In this section, the study will be carried out on methods proposed when introducing new challenging datasets for individual analysis purposes, as in the related work by Rendón-Segador et al. (2023). As part of our comprehensive research efforts, the study will also examine a method by Yi et al. (2020) proposed for a dataset built on his own for terrorism-type attacks.

Sultani et al. (2018) introduced the UCF-Crime collection which includes 14 types of violence classes and a DMIL ranking model. The NTU CCTV Fights dataset was presented in 2019 in Perez et al. (2019). It contains long violent videos that are labeled at frame level. In Degardin and Proença (2020), they released the UBI-Fights dataset in 2020 and proposed an architecture based on the Gaussian mixture model for anomaly detection. In Wu et al. (2020), the authors also released the XD-Violence dataset in the same year, which includes sound and seven classes. They presented a method that exploits visual and audio feature extractors to detect short- and long-range temporal relationships using graph neural networks. Ramachandra and Jones (2019) introduced the Street Scene dataset in the same year. It contains 17 anomaly types and introduces two variations of a novel baseline video anomaly detection algorithm based on handcrafted features and a distance function to detect anomalies.

Dataset	**Feature extractor**	**Classifier**	**Result**
UCF-Crime	C3D	FNN	75 AUC
NTU CCTV Fights	I3D/C3D	CNN or LSTM or SVM	75.9 f-score 79.5 mAP
UBI-Fights	C3D	FNN + GMM	82 AUC
XD-Violence	C3D & VGGish	GCN	78 AP
Street scene	Optical flow or foreground mask	Distance function with threshold	Track based and region-based ROC

Table 1.7. *Comparative study of methods in individual behavior scene analysis*

These methods establish a baseline in state of the art (SoTA) which is then challenged by other approaches regarding functionalities and performance. In Table 1.7, a deep study of used methods is outlined. The method presented by Sultani et al. (2018) proposed a C3D and MLP in a multi-instance learning framework with a performance rate of 75% AUC on UCF-Crime. The method presented by Perez et al. (2019) proposed an I3D/C3D with CNN/LSTM or an SVM classifier with a performance rate of 75.9% F-measure and a mean Average Precision (mAP) of 79.5%. The method presented by Degardin and Proença (2020) proposed a C3D /FFNN and GMM overpredicted scores in negative bags and positive bags separately, which allows for smoothing the scores with a performance rate of 82% AUC on UBI-Fights. The method presented by Wu et al. (2020) proposed a C3D for visual feature extraction and VGGish for audio feature extraction and a GCN with a performance rate of 78% mAP. The method presented by Ramachandra and Jones (2019) proposed the optical flow or foreground mask for feature extraction and a distance function with a threshold as a classifier which was evaluated on the following curves: track-based and region-based ROC.

Additionally, some research on audio, image and video analyses has occurred for terrorism detection but was not that explicitly developed due to a dataset developed on its own. For instance, in the method presented by Yi et al. (2020), the authors study terrorism detection by analyzing audio and images on a dataset where image and audio data were not correlated. They proposed a framework that detects file name detection, audio and image detection, as mentioned in Table 1.8.

<table>
<tr><th></th><th>File name</th><th>Audio</th><th>Image</th></tr>
<tr><td>Feature extractor</td><td>Word2vec</td><td>Handcrafted features (mfcc; chroma, mel, contrast, tonnetz)</td><td rowspan="2">Faster R-CNN, single shot multibox detector (SSD), and region-based fully convolutional network (RFCN) for video recognition</td></tr>
<tr><td>Classifier</td><td>Word similarity</td><td>Logistic regression, decision tree, SVM, XGboosting, light GBM</td></tr>
</table>

Table 1.8. *Methods for terrorism detection in file names, audio and image analysis*

The majority of these methods use deep learning-based methods thanks to their strong ability to generalize to different contextualized anomalies until improvement of the learning framework, schema and loss or simply the model architecture with some CNN, RNN or attention mechanism component. Others use some handcrafted feature extraction methods and shallow models to classify them.

1.2.5. *General human threat-centric urban situation analysis*

The areas of application identified include traffic signals and main junctions, residential areas, crowd-pulling meetings, festivals as part of religious institutions and inside office buildings. Many existing solutions are tailored specifically to these contexts. However, crowd analysis is often the most challenging due to the need to accurately identify a wide range of actions, behaviors and movements. The analysis of crowd scenes is a well-established area of research, while traffic and the environment are less studied, resulting in public datasets built by the researchers themselves in these two fields. The analysis of individual behavior is presented as the new challenge for urban video surveillance, and new challenging datasets have been proposed. To cover all the problems at the same time, we are looking for a public dataset available to compare our method with the state of the art.

Name	Detail	Provider	Source/Localization	Use	Label	Anomaly	Comment
CUHK Crowd	474 videos; 215 crowd scenes. Various resolutions. 8 classes: very mixed pedestrian traffic, crowds walking in well-organized places, in poorly organized places, merging, splitting, crossing in opposite directions, disturbed escalator traffic, smooth escalator traffic	The Chinese University of Hong Kong	Internet (Pond5 and GettyImages)	Behavior	Yes	Run, throw, new object	Less interesting application case because the training data is not annotated with interesting labels. The camera angles are interesting for generalization tests
Avenue	37 videos: 16 training and 21 testing. Training set: low amount of anomalies and normal situations rarely appear. Test set: normal/abnormal events	The Chinese University of Hong Kong	Surveillance cameras of an avenue of the CUHK campus	Behavior	Yes	Running, throwing an object, new object, pushing a bike, strolling, wrong direction	Field application less interesting in terms of anomalies. Little usable data. Interesting camera view, but irrelevant content
ShanghaiTech Campus	13 scenes with complex and varied lighting and camera angle conditions. 130 abnormal events	Shanghaitech Vision and Intelligent Perception (SVIP) Lab, ShanghaiTech University	ShanghaiTech University campus surveillance cameras	Behavior	Yes	Sudden movement (chase, fight), bicycle, baby stroller, path crossing	Data more important than UCSD, CUHK and Subway according to the provider because they come from different cameras
ShangaiTech Part A&B	Part A: varied resolution, part B: resolution 768*1024	X	X	Counting	X	X	Academic research for crowd modeling. Not interesting for implementing a solution
UCSD PED1/2	PED1: 34 training videos and 36 test videos. Resolution 238×158, view: stationary camera. PED2: 16 training videos and 12 test videos. Resolution 238×158, view: stationary camera	Statistical Visual Computing Lab (SVCL), University of California San Diego	Stationary surveillance cameras mounted high up, with a view of pedestrian walkways	Behavior	Yes	Circulation of non-pedestrian entities (bicycles, carts) and abnormal dynamics of pedestrians (crossing paths)	Field application less interesting in terms of anomalies
UMN	3 scenes (campus lawn, indoor and square) and 11 videos. UMN focuses on a single behavior "panic movements"	Project "Monitoring Human Activity", Artificial Intelligence, Robotics and Vision (AIRV) Laboratory, University of Minnesota	X	Behavior	Yes	Crowds running in one direction, a crowd of people dispersing from a central point	Field application less interesting in terms of anomalies
WorldExpo'10	2650 videos, 245 scenes, Resolution: 576*720	Shanghai Jiao Tong University and The Chinese University of Hong Kong	Shanghai 2010 WorldExpo in China (road, queue, square)	Counting	X	X	Academic research for crowd modeling. Not interesting for implementing a solution
Mall	640x480 resolution, more than 60,000 pedestrians have been labeled in 2000 video frames. Annotations of the head position of each pedestrian in all images	The Chinese University of Hong Kong	From a publicly accessible webcam	Counting/ profiling	X	X	Academic research for crowd modeling. Not interesting for implementing a solution
UCF_CC	38 videos, 38 scenes, Resolution 480*360	University of Central Florida, Center for Research in Computer Vision	Internet (FLICKR)	Counting	X	X	Academic research for crowd modeling. Not interesting for implementing a solution
UCF-Crime	1900 videos, 128 hours of videos	University of Central Florida, Center for Research in Computer Vision	Original surveillance video footage from internet (YouTube, LiveLeak) around the world	Behavior	Yes	Abuse, arrest, arson, assault, accident, burglary, fight, robbery	Most complex dataset (complex behavior of scenes, view, angles, resolution, quality, varied color), great generalization capacity
UBI-Fights	1000 videos, 80 hours of videos, 640*360 resolutions	Universidade da Beira Interior	Mozambique	Behavior	Yes	Fight	Dataset that identifies attacks, a dataset that can be interesting at the beginning
XD-Violence	4754 videos, 2405 abnormal and 2349 normal, audio-visual dataset, from films and YouTube (91 films)	Xidian University	From the internet (films and YouTube)	Behavior	Yes	Abuse, car accident, explosion, assault, shooting, riot	May introduce a bias to generalization and solution is not applicable in most video surveillance systems

Table 1.9. *Comparative study of datasets*

1.2.5.1. *Existing public dataset review and analysis*

Establishing this proper dataset in Europe is complicated due to the GDPR. So, finding a suitable publicly available dataset was crucial. Moreover, as shown in Table 1.9, by analyzing providers, the dataset was mainly in China and the USA. We create three categories of crowd monitoring including localization, behavior and counting. Detecting violent scenes is of significant interest in smart cities for public safety. We divide it into two big types of analyses: complex urban (crowded) scene analysis and traditional urban (uncrowded) scene analysis. The kind of anomaly to detect is important and useful for surveillance applications. It will conduct a better analysis and provide relevant information to first responders or the control room.

A deeper analysis of anomalies present in datasets is shown in Table 1.10.

To conclude, the dataset that stands out in terms of the anomalies contained within is the UCF-Crime dataset.

1.2.5.2. *State of the art on UCF-Crime*

Knowing the limits of this dataset and learning, some related works have been done by categorizing unsupervised learning or weekly supervised learning, encoder-agnostic learning or encoder-based learning, spatial or temporal coherence learning, attention-based learning, or not. In our related work, we are going to overcome all these systematic categorizations by considering all of it as overall machine learning methods and detail these categories as extended modules to push the analysis further. The UCF-Crime dataset is introduced in Sultani et al. (2018), and work is done with the 3D Convolutional Neural Network (C3D) (Tran et al. 2014). Spatial and temporal features are extracted using the CNN and then classified with a multi-layer perceptron. And where only the label is known for the bag containing 32 instances, Multi-Instance Learning (MIL) is used. The algorithm will then be able to detect the segment-level label despite having only a video-level label dataset. Anala et al. (2019) show the use of an enhanced 2D CNN called VGG16 (Simonyan and Zisserman 2014) to extract spatial features and a Long Short-Term Memory (LSTM) as a sequence learner, thereby providing better accuracy, but the AUC is not specified. In Li et al. (2021), spatial and temporal features are extracted using a generative model. This was done by deploying an autoencoder learner with the construction error to take advantage of the feature present in latent space, which in turn outperformed many datasets but not with this chosen dataset. In Aberkane and Elarbi (2019), the first presented 3D CNN (Tran et al. 2014) feature extractor is used with an RL model. A Deep Q learning model is used to detect anomalies instead of a supervised learning model. This approach reuses a MIL-classifier to detect abnormal segments and uses its output as a feedback for the reinforcement algorithm rather than consistent RL, and it brings a 3% performance increase. Although the approach (Aberkane and Elarbi 2019) is not relevant in this context, using RL for anomaly detection in video surveillance helps take the research forward. In Lv et al. (2020), they extract features from a

	CUHK Crowd	Avenue	ShangaiTech Campus	UCSD Ped1/2	UMN	UCF-Crime	UBI-Fights	XD Violence
Vandalism						X		
Shooting						X		X
Explosion						X		X
Riot								X
Agression						X		X
Shoplifting, stealing, robbery						X		
Burglary						X		
Accident						X		
Fire						X		
Arrest						X		X
Abuse						X		
Bike/cart			X	X				
Figthing			X			X	X	X
Stroller			X					
Bad direction		X	X	X	X			
Pushing a bike		X						
New object	X	X						
Throwing	X	X						
Run	X	X	X		X			

Table 1.10. *Comparative study of anomalies in datasets*

Temporal Segment Network (TSN) (Wang et al. 2016) and perform MIL using margin loss as in Sultani et al. (2018) and compute this loss also on the dynamic variations of the segments obtained by their cosine similarity and also perform this new MIL loss to a modified dataset containing simulated noises and pseudoanomalies created by merging normal and abnormal segments. This approach provides a real improvement compared to the first baseline approach, with a gain of 10%. In Feng et al. (2021), they extract features from a Two-Stream Inflated 3D ConvNET (I3D) (Carreira and Zisserman 2017) and perform, first, MIL using margin loss in Sultani et al. (2018) and, second, encoder learning with self-guided attention mechanisms in CNN to learn a task-specific encoder in order to eliminate the domain gap from the pre-trained parameters. This approach provides an improvement compared to the first baseline approach, with a gain of 7%, and shows the impact of the encoder. In Tian et al. (2021), they extract features from a Two-Stream Inflated 3D ConvNET (I3D) (Carreira and Zisserman 2017) and perform, first, MIL on top-K snippet by considering multiscale temporal dependencies also called magnitude learning using margin loss in Zhou et al. (2022) and, do RTFM enabled snippet classifier. This approach provides an improvement compared to the first baseline approach, with a gain of 9% without doing encoder fine tuning or an augmented dataset, and purely outperforms other MIL-based methods. In Chen et al. (2022), they extract features from a Two-Stream Inflated 3D ConvNET (I3D) (Carreira and Zisserman 2017) and perform the Tian et al. (2021) method by adding a feature amplification mechanism and modify the existing temporal coherence learning by a Glance block and by adding a focus block for spatial feature learning both inspired by attention mechanisms embedded in transformers. This approach provides an improvement compared to the first baseline approach, with a gain of 11% without doing encoder fine-tuning or an augmented dataset and purely outperforms other MIL-based methods by integrating an attention mechanism as for (Feng et al. 2021) feature learning and for (Tian et al. 2021) for temporal coherence learning inspired from the first by transformers and the second of CNN-based attention.

First, in Table 1.11, the analysis of feature extraction methods is pushed further before going deeper in the anomaly detection framework for classification.

As we can see, many C3D (Tran et al. 2014) feature extraction methods were used in their proposals, which make a SoTA FE method to use. Nevertheless, some other approaches exceed in terms of performance such as I3D (Carreira and Zisserman 2017) or TSN (Wang et al. 2016) FE methods due to their highest capabilities to model spatial–temporal relationships as in I3D (Carreira and Zisserman 2017) by analyzing optical flow or by capturing better temporal relationships in TSN.

Now, to get a deeper understanding of the anomaly detection framework, details are provided in Table 1.12.

Ref.	Year	Feature Extractor	Result	Gap
Sultani et al. (2018)	2018	C3D (Tran et al. 2014)	75.41	-
Anala et al. (2019)	2019	VGG16 (Simonyan and Zisserman 2014)+LSTM	85	-
Li et al. (2021)	2021	C3D (Tran et al. 2014)	72.9	-
Aberkane and Elarbi (2019)	2019	C3D (Tran et al. 2014)	78.20	-
Lv et al. (2020)	2020	I3D (Carreira and Zisserman 2017)/TSN (Wang et al. 2016)	74.18/ 84.44	10.26
Feng et al. (2021)	2021	C3D (Tran et al. 2014)/I3D (Carreira and Zisserman 2017)	81.40/ 82.30	0.9
Tian et al. (2021)	2021	C3D (Tran et al. 2014)/I3D (Carreira and Zisserman 2017)	83.28/ 84.30	1.02
Chen et al. (2022)	2022	I3D (Carreira and Zisserman 2017)/ SWIN Liu et al. (2021)	86.98/ 86.67	0.31

Table 1.11. *Comparison study of feature extractors*

In order to obtain a holistic understanding of the methods and contributions presented, we have meticulously compiled and presented them in a clear and concise manner in Table 1.13.

In the study by Anala et al. (2019), they treat an entire sequence as one element in the spatial representation and in the study by Li et al. (2021), they focus on spatial representation due to complete blind learning. In the study by Sultani et al. (2018), they use a very small weight but it has an impact on the result, comparing without this temporal term in loss function and in the study by Aberkane and Elarbi (2019), they use the first method and try to break the correlation between segments with experience replay which arranges segments in a random order for learning. In the study by Lv et al. (2020), they deal with spatial and temporal coherence, respectively, called semantic and dynamic variations in their approach. In the study by Feng et al. (2021), they deal with spatial coherence only and go deeper into spatial coherence until fine-tuning the encoder to a specific task. In the study by Tian et al. (2021), they deal with spatial and temporal coherence at the same time by optimizing on spatial domain, considering temporal dependencies of segments. In the study by (Chen et al. 2022), as in the previous one, they deal with spatial and temporal coherence at the same time by optimizing the spatial domain, considering temporal dependencies of segments.

In summary, certain methods are only effective in analyzing spatial data while others are only useful in analyzing temporal data. However, none of these methods allow for anticipating the future by taking into account the impact of past states on a specific situation. Furthermore, when it comes to evaluating the performance of multiple instance learning, incorporating an attention mechanism significantly improves its accuracy.

Ref.	ML Method	Loss Function	Framework Goal	Training and Decision Level	Comment	Result
Sultani et al. (2018)	Semi-supervised contrastive learning/multi-instance learning	Custom hinge loss	Maximize distance between abnormal and normal segments	Video level learning label and level prediction	Coherent MIL framework for decision-level prediction	75.41% AUC
Anala et al. (2019)	Supervised Sequence learning	Cross Binary Entropy	Action recognition to detect anomaly at video level	Training and Video level prediction	Not suitable with real-time compliance strategy	85% ACC (reduced dataset)
Li et al. (2021)	Unsupervised representation learning and clustering filtering	Reconstruction error and Kullback–Leibler (KL) divergence based clustering error	Learn normal representation and cluster normal and abnormal segments	Video level learning Label and level prediction	Decision-level prediction with a not suitable inductive bias and manual threshold setting	72.9% AUC
Aberkane and Elarbi (2019)	MIL and supervised reinforcement learning	Zhou et al. (2022) + MSE between estimated and Q value and real Q value built on rewards	Zhou et al. (2022) and discriminate abnormal and normal segments	Zhou et al. (2022) + Label level learning Label and level prediction	Decision-level prediction based on uncertain ground truth	78.2% AUC
Lv et al. (2020)	MIL and similarity learning	margin loss of semantic (Euclidean) and dynamic variation (cosines) with sparsity constraints	Maximize distance between abnormal and normal segments on original and modified data (noise and pseudo anomaly inserted)	Video level learning Label level prediction	Coherent MIL framework for decision-level prediction	85.38% AUC (augmented dataset)
Feng et al. (2021)	MIL and encoder-based supervised learning	Hinge loss for MIL with sparsity constraint and CE for encoder learning	Maximize distances between abnormal and normal segments and use its pseudo labels to retrain the encoder with an attention mechanism on spatial features.	Video level learning Label level prediction	Coherent MIL framework for decision-level prediction	82.3% AUC (encoder retained) C3D/I3D
Tian et al. (2021)	MIL of top K "magnitude learning" with multiscale temporal network and RTFM-enabled snippet classifier	Hinge loss for MIL and CE over chosen snippets	Maximize distance between abnormal and normal top K largest snippets according to its local and global temporal dependencies based on the pyramid of dilated convolutions and a temporal self-attention module and proceed to snippet classifier learning	Video level learning Label level prediction	Coherent MIL framework for decision-level prediction	84.30% AUC C3D /I3D
Chen et al. (2022)	RTFM improved by feature amplify mechanisms, and use its own temporal coherence learning called Global block and by adding a spatial feature learning called Focus block, both inspired from attention mechanism, first from transformers and second for CNN	Hinge loss for MIL and CE over chosen snippets	Maximize distances between abnormal and normal top K largest snippets according to a separate path containing a feature amplify mechanism, and its attention mechanisms for temporal coherence and spatial features and proceed to snippet classifier learning	Video level learning Label level prediction	Coherent MIL framework for decision-level prediction	86.98% AUC I3D/VideoSwin RGB

Table 1.12. *Comparison study of anomaly detection framework*

Ref.	Coherence Spatial	Coherence Temporal	Pixel-Level Analysis	Attention Mechanism	Anticipation
Sultani et al. (2018)	Yes	Yes	No	No	No
Anala et al. (2019)	Yes	No	No	No	No
Li et al. (2021)	Yes	No	No	No	No
Aberkane and Elarbi (2019)	Yes	No	No	No	No
Lv et al. (2020)	Yes	Yes	No	No	No
Feng et al. (2021)	Yes	No	Yes	Yes	No
Tian et al. (2021)	Yes	Yes	No	Yes	No
Chen et al. (2022)	Yes	Yes	Yes	Yes	No

Table 1.13. *Comparative study of extended cognitive modules*

1.3. System architecture for video surveillance

In the field of video analytics, the ground truth is provided by the user for reliability purposes, and the Quality of Prediction (QoP) metric remains significant. Similarly, in intelligent network controls, involving the end-user enables a complete system view for effective control. QoE is the solution that provides the end-user's ground truth. Therefore, our study focuses on optimizing this metric to enhance the system's overall performance. First, the network architecture will be studied to improve the QoE of video surveillance systems. In addition to existing methods, we will explore how computing infrastructure can further enhance it.

1.3.1. *Network architecture*

These studies were generally divided into three categories and could be regrouped by how to:

– model the QoE;

– identify which parameters impact the QoE most in order to enhance these essential parameters;

– self-remediate to ensure QoE.

The remediation work consists of improving the QoE by acting directly on the concerning environment, i.e. the network. The network is assessed and enhanced progressively in order to improve user satisfaction. The referential of the study is clearly the network, "The network for the network".

These studies focused on complex architectures such as Vehicular Ad Hoc Networks (VANET), Wireless Camera Networks (WCN) and Unmanned Aerial Vehicles (UAV) for video streaming applications, with a special emphasis on video surveillance. These studies involved a lot of variability in parameters and impact factors and were conducted using the most sophisticated architecture available.

Next-generation network challenges include adapting to the increasing demand for higher bandwidth, ensuring compatibility with emerging technologies, addressing security risks posed by cyber threats and optimizing network performance in the face of growing complexity. Ongoing research and development is crucial to meeting these challenges and ensuring that next-generation networks are reliable, efficient and secure.

There are studies that aim to enhance subjective service performance for video surveillance networks. These studies specifically focus on complex video surveillance architectures such as WCNs, mobile networks and specific architectures such asVANETs and UAVs due to their potential uses in video surveillance. However, such complex architectures pose new challenges, such as resource allocation, dynamic topologies and how to measure performance.

Ref.	Network Architecture	QoE	Goal	Solution	Automation
Wu et al. (2012)	Wireless Camera Network (WCN) over 4G mobile networks, including WiMAX and LTE	Aggregated utility values of all DL mobile stations, subject to a finite amount of wireless resources	Effectively scheduling and allocating resources to the UL videos from all camera stations (CS) + Maximize the distributing efficiency of the DL videos to all MSs	CrossLayer QoE-Driven Scheduling	Amount of OFDM Slot allocation for this video and how many video layers are allowed to uplink + scheduling video packets in each OFDMA frame
Wang et al. (2018a)	Mobile networks	Video Mean Opinion Score (vMOS)	Investigating the relationship between 11 features (i.e. QoS parameters) and 1 scoring factor vMOS on mobile network "SVGOP" of Huewai	Network parameter-based vMOS computing by a combination of K-mean clustering method and logistic regression method. + Identify a small set of QoS parameters that play an important role in determining vMOS (initial buffering latency, stalling times and stalling ratio)	-
Naveed and Qazi (2019)	4G-LTE for Unmanned Aerial Vehicle (UAV) networks	Peak-signal-to-noise-ratio (PSNR) and Structural-Similarity Index (SSIM)	Comparative analysis between two latest codec schemes, i.e. H.264 and H.265 (HEVC) in video streaming + Analyze the effect of different factors on QoE of video streaming across the network according to 3GPP R4-092042 specification	H.265 better than H.264 under different circumstances (Line of Sight to Non-Line of Sight threshold (LoS2NLoS), varying macroENB sites, varying homeUEs per homeENB ratio and varying internal wall-loss.)	-
Quadros et al. (2015)	Vehicular ad hoc network (VANET)	Average video distortion (and Link Quality which is not considered as QoE)	Improve video dissemination in V2V scenarios + Improve the transmission of on-road videos for medium distances (e.g. 1–5 km) as required for many disaster, surveillance, and even for entertainment in multimedia smart city services.	Employs QoE-indicators criterion that supports the selection of the best next hop and switches to other routes as soon as lower quality is identified via QOe-Driven and LInk-qualiTy rEceiver-based (QOALITE), which is aware of transmission routes and node failures.	Relay node selection (the selection of the best next hop and switches to other routes)
Quadros et al. (2016)	Vehicular Network	SSIM and MOS (both QoE)	Improve video dissemination in dynamic topology, V2V scenarios	Cross-layer QOe-driven REceiver-based (QORE) mechanism (where no information about neighbors but neighbors dispute to forward data), which is modularly coupled to Distance-based Statistical Routing Protocols (SRPs) to offer QoE-aware and video-related parameters for the relay node selection and backbone maintenance	Selection of the best next hops, changing to other nodes

Table 1.14. *Existing studies on cross-related work between QoE and video surveillance*

As shown in Table 1.14, various works have been carried out in the field of video surveillance to guarantee the quality of experience in next-generation networks. In Wang et al. (2018a), the work consists of measuring performance and finding the best way to model user satisfaction, i.e. the video Mean Opinion Score (MOS). Other work involves improving video broadcast performance by formulating the problem as one of resource allocation, as in Wu et al. (2012) with slot and frame allocation, or as in Quadros et al. (2015, 2016) by selecting the best relay node. Studies by Wu et al. (2012) and Quadros et al. (2015, 2016) have the advantage of being adaptable to factors influencing performance and of providing an adaptable and reliable network to certain factors. In Naveed and Qazi (2019), the study consists of selecting the best codec schemes according to the 3GPP R4-092042 specification in order to establish a static configuration for video surveillance on a UAV network. QoE modeling, configuration selection in a stationary environment or self-adapting networks for dynamic changes are at the heart of research for next-generation video surveillance systems and are due to the inherent complexity of their network architectures. The way in which user satisfaction has been modeled attracts our attention. In all these approaches, user satisfaction takes a different form and is formulated as a subjective evaluation measure (MOS), as an objective evaluation measure (PSNR, SSIM or average video distortion) or as a utility function with a budget constraint. In Wu et al. (2012) and Quadros et al. (2015, 2016), these solutions provide an adaptive solution for improving the quality of experience of the video surveillance network by observing certain measurements and controlling certain parameters. For all these three approaches, the observation space is the network, and the control space is the network, which enables self-management of the network.

1.3.2. *Computing infrastructure*

Over the years, computing infrastructure has evolved significantly in video surveillance systems. With the advent of cloud computing, fog computing and edge computing, it has become possible to distribute computing resources and processing power more efficiently. This has led to increased scalability, reliability and cost-effectiveness in video surveillance systems. Cloud computing, in particular, has made it possible to store and manage large amounts of video data securely and access it from anywhere, while fog computing and edge computing have enabled real-time processing and analysis of video data at the edge of the network. Overall, these computing paradigms have revolutionized the way video surveillance systems are designed and deployed, making them more powerful and effective than ever before. Cloud computing, fog computing and edge computing are all computing paradigms that enable distributed computing and processing. Cloud computing involves using a network of remote servers to store, manage and process data, while fog computing focuses on bringing computing resources closer to where they are needed, typically at the edge of the network. Edge computing involves processing data locally on devices at the edge of the network, rather than sending it to a remote server for processing. Each of these paradigms has its own unique strengths and drawbacks and can be used in different ways to meet specific computing needs.

As shown in Table 1.15, in Rodríguez-Silva et al. (2012), the authors proposed a cloud-based video surveillance system that collects the multimedia streams generated by surveillance cameras and stores them in a cloud. Knowing the weaknesses of the cloud computing approach in terms of security or real-time application, they optimize transmission according to the state of the network. In Chen et al. (2016), the authors proposed a fog-based video surveillance system for traffic monitoring. It allows for multi-target tracking using a single target tracking algorithm where the tracking tasks for each vehicle in video frames are dispatched to on-site available fog nodes. Fog nodes are all on-site smart devices such as smart tablets, smartphones, computers in police cars and other on-site devices with computation capabilities to process collected data from end-users and feed the results back in a timely manner. Here, the drone acts as a camera sensor in the sky, monitoring the areas of interest. In Wan et al. (2021), the authors suggest using edge computing for video pre-processing in order to eliminate redundant frames. This would allow for the possibility of migrating part or all of the video processing tasks to the edge, ultimately reducing the computing, storage and network bandwidth requirements of the cloud center. In Sun et al. (2017), the authors propose a video usefulness model for large-scale video surveillance systems based on edge computing and cloud computing. Thanks to edge computing, video data can be pre-processed, and certain faults can be localized to reduce the load on the network and optimize video data storage in the cloud. In addition, deep-level usefulness evaluation can be carried out in the cloud. In Zhang et al. (2016), the authors propose EVAPS (Edge Video Analysis for Public Safety), which distributes the computing workload in both the edge nodes and the cloud in an optimized way. EVPAS is able to eliminate unnecessary data transmission over it and save energy for edge devices, i.e. cameras.

Computing Infrastructure	Purpose	Ref.
Cloud	Storage, collection (IoT services), processing big data	Rodríguez-Silva et al. (2012) Sun et al. (2017) Zhang et al. (2016)
Fog	On-site smart devices for the computational task in an intermediate localization distance	Chen et al. (2016)
Edge	Energy and time efficiency, security	Wan et al. (2021) Sun et al. (2017) Zhang et al. (2016)
Hybrid cloud and edge	Real-time analysis (edge) and storage (Cloud)	Sun et al. (2017) Zhang et al. (2016)

Table 1.15. *Existing studies on cross-related work between computing infrastructure and video surveillance*

To summarize, the fog-based approach could be complicated to deploy in real-world scenarios due to the use of ubiquitous computing, which focuses on embedding computing devices into everyday objects and environments, and where access to ubiquitous devices could be complicated due to privacy concerns. To better address the obstacles associated with edge and cloud computing, it is worth exploring hybrid approaches that optimize their collaboration. Such studies could prove more intriguing by focusing on each approach separately and finding ways to enhance their cooperation. In fact, Zhang et al. (2016) and Sun et al. (2017) use cloud- and edge-based computing together; both optimize their cooperation by filtering useless data at edge boundaries to transmit and store it on the cloud.

1.4. Analytics and architecture: studies and reflections

In video surveillance, as shown in previous sections, analytics and architecture were studied separately to cope with their respective problems. In some cases, thinking together could improve the performance of the overall system. As a preliminary study, we look first for threats in cyber and physical space and their impact in the other space; second, the impact of data quality towards data-driven analytics solutions; and third, to optimize a system regarding an overall system quality assessment.

1.4.1. *Threats: from cyber-to-physical space or physical-to-cyber space?*

Cyber refers to the architecture, and physical refers to the content analyzed by the analytics component. Some threats from cyberspace could have negative impacts on physical space by deteriorating or dumbing-down analytics, and some threats from physical space, such as vandalism on CCTV infrastructure caught by the analytics, could impact the architecture negatively. Both could cause dramatic situations for public safety.

Cyber-to-physical

Cyberspace has an impact on physical space when it comes to urban threats; this can be due to network defiance or cyber attacks, which differentiate them by unintentional or intentional failure. Here, intentional failure will be studied, i.e. cyberattacks to highlight urban threats due to intended intention. Data privacy is crucial when it comes to cyber attacks. With the increasing number of cyber attacks, protecting our personal and sensitive information has become more important than ever. Hackers are constantly looking for ways to steal data and use it for their own gain. It is essential to take all necessary measures to safeguard our information and prevent cyber attacks from happening. By being mindful of our online activity and taking precautions such as using strong passwords and avoiding suspicious emails, we can help ensure our data remains private and secure. When it comes to cyber attacks, vulnerability can be divided into: availability, confidentiality and integrity. Some attacks that have occurred in recent years are listed in the following study (Vennam et al. 2021). In Table 1.16, some of these attacks on CCTV are mentioned by their data vulnerability.

Vulnerability	Ref.	Attack type	Consequence
Confidentiality and availability (storage or real time)	CyberScoop (2017)	Cerber ransomware	Stored data access and availability control
	vdoo (n.d.)	Command injection and control attack	Real-time data access and availability control
Availability (real time) and integrity	CSO online (n.d.), Bosch (n.d.)	DDoS attack by botnets and Buffer overflow	Device performance degradation + corrupting or overwriting data. Both availability and integrity impact

Table 1.16. *Referred attacks on video surveillance system*

Some of these attacks have to exploit data confidentiality coupled with availability vulnerabilities, and others exploit availability with integrity vulnerabilities. These attacks occurred in recent years and must garner attention in video surveillance deployment and design. Giving attention to the analytics component in recent years, data poisoning attracted the attention of lots of researchers in the AI community to work in the field of Trustworthy AI. Data poisoning is an attack that dumps the analytic component. This study has been appropriated to the AI community as visual-based attacks; it differentiated this to network-based attacks, which is a topic that tackled the computer network community. In our study, we will not oppose these but will integrate visual-based attacks as data engineering injection attacks among network-based attacks. On the other point, this refers to advanced data poisoning. Some network-based attacks have been mentioned; nevertheless, in the literature on AI, visual-based attacks receive greater attention because they can deceive the model. Obviously, visual-based attack approaches must be initiated by network one, which interferes with the data integrity by injecting data with the aim of corrupting the learned model. When someone manipulates the training data for a machine learning model, it is known as a data poisoning or model poisoning attack. It is considered an integrity attack since it can interfere with the model's capacity to provide accurate predictions. Among visual data poisoning, two of them are classified in Table 1.17.

Ref.	Breach level	Method	Goal
Shafahi et al. (2018)	Model learning	Clean-label poisoning attacks on neural network	Proceed to data poisoning without having labeling control and interfering with overall model performance which could lead to performance drop and ease detection
Bagdasaryan et al. (2018)	Distributed learning strategy	New model-poisoning methodology based on model replacement	Proceed to global model poisoning by introducing an attacker in a single round of federated learning

Table 1.17. *Two cases of visual data poisoning on AI-based video analytics*

Among these studies, visual data poisoning could occur in model learning or through its distributed learning strategy.

Physical-to-cyber

In any video surveillance system, the most important hardware components are the cameras, known as the eye of the video surveillance system. They play a critical role in safeguarding tangible assets and ensuring safety and security. Protecting security cameras should be a top priority, and this includes prioritizing both physical and data security measures. Vandalism of CCTV camera infrastructure is a physical threat that impacts cyberspace, which will not convey content to the expected authorities. Among many examples, Chicago CCTV cameras were often vandalized. As mentioned in CES Complete (n.d.), when someone intentionally damages a security camera or tampers with its settings, it can compromise the camera's ability to capture accurate footage and deter criminal activity. This can be especially harmful to businesses, as it can lead to financial losses and increased vulnerability to crime. Vandalism may also target cameras for the sake of causing damage and disruption.

As we can see, physical threats are little studied, even if they can lead to dramatic situations, because the research is done at the limit of the possible. Only a physical solution is possible to answer this problem, while the scope of possibilities opens up when it comes to software.

1.4.2. *Video quality impact on video surveillance: from monitoring to task-specific analytics*

Video quality is a crucial factor when it comes to video surveillance for specific tasks. The quality of the video can have a significant impact on the accuracy of video analysis and recognition. In order to ensure accurate detection and identification of objects or people, it is essential to have high-quality video.

In Korshunov and Ooi (2011), they conducted experiments that show the effect of video quality degradation as blockiness on several typical examples of face analysis algorithms. The surprising finding of this article is that the tested algorithms show very high tolerance to large reductions in video quality, which amounts to at least 10 times less video bitrate than video conventionally encoded for human vision. They argued that algorithm-oriented video quality metrics need to be developed. In Sellahewa and Jassim (2010), they investigated the challenges of face recognition in the presence of extreme variation in lighting conditions. From the Luminance Quality (LQ) Index, they develop global and region quality-based adaptive illumination normalization procedures. They demonstrated the effectiveness of the proposed image quality-based illumination normalization schemes in face recognition compared to the traditional approach. In Yang et al. (2018), they demonstrated that it is not the image quality itself that is the most important, but rather that the quality of the images in the training set should have similar quality as those in the testing set. They proposed a deep neural network first to classify images based on their quality classes and then to design a separate face detector and recognizer for images in each quality class. They present experimental results to show that their quality classified framework can accurately classify images based on the type and severity of image degradation and can significantly boost the performances of state-of-the-art face detectors and recognizers in dealing with image datasets containing mixed-quality images.

As shown in Table 1.18, the works of Korshunov and Ooi (2011) focuses on how to properly assess video quality so as not to degrade a specific video surveillance task, the work of Sellahewa and Jassim (2010) focuses on how to improve degraded data before ingesting it into the analysis engine, and the work of Yang et al. (2018) focuses on how the analysis engine will handle degraded data quality. These approaches have been introduced in order of the complexity of the analysis engine's capabilities.

Ref.	Key point
Korshunov and Ooi (2011)	Link between video quality and video surveillance-specific task performance. + Existing human assessment metrics are not representative of assessing video surveillance-specific task performance
Sellahewa and Jassim (2010)	Present adaptive approaches to face recognition to overcome the adverse effects of varying lighting conditions
Yang et al. (2018)	Training and testing data should have the same quality to preserve performance level. + Present a quality-aware learning and inferring framework for face detection and recognition

Table 1.18. *Studies on video quality impact on video surveillance analytics tasks*

1.4.3. *End-to-end measurement: from traditional QoE to task-specific QoE*

Task-specific QoE refers to the quality of experience (QoE) that is tailored to a particular task or application. In the context of video analysis and recognition in surveillance tasks, task-specific QoE would refer to the level of video quality required to accurately perform these tasks. Task-specific QoE in video surveillance is poorly explored but the study by Abar et al. (2022) is by far the solution with the most similar objective to ours; it improves video surveillance decision support, its resources, false detection of abnormal behavior and definition of an optimal perimeter, but it has no experimental section and has some limitations; the accuracy of abnormal behavior detection is specific to a university case and could be improved in the case of a multi-camera configuration. The video analysis component has not been improved, but performance has been enhanced by context pruning and needs more video analysis components to enable it.

In video surveillance, traditional QoE offers self-remediation on the network, and task-specific VQ studies propose to increase data quality or to deal with a variety of data quality to ensure final task performance. What if we want to apply for a general end-to-end measure, i.e. a task-specific QoE in order to deploy it in any network infrastructure/architecture (different noises and disturbances), and we do not have the ability to operate on it or identify the exact disturbance in order to correct it before ingesting this data? We need to preserve a component that is intelligent, reliable and impervious to any data degradation that takes place in any network configuration without the possibility of optimizing it.

Ref.	Goal	Task Specific QoE	Solution	Video surveillance specific task
Abar et al. (2022)	Proposed a distributed intelligent surveillance system based on smart cameras by enhancing the QoE operator side by minimizing overall resource consumption and false detection and optimizing the tasks assigned to the cameras	1) Resource availability constraints 2) False detection of suspicious behavior 3) Define an optimal perimeter for intrusion detection	1) Just back up the necessary records that identify a kind of suspicious behavior or switch off unnecessary cameras to preserve power 2) False detection of suspicious behavior: make the algorithm run exactly where it is needed to avoid false detection, i.e. predicting an anomaly where it is not necessary or missing an intrusion that took place 3) Define an optimal perimeter for intrusion detection: depending on the route that will be chosen, the cameras take the relay of the anomalous behavior	Intrusion detection (recognition, tracking + detection) over more than one camera

Table 1.19. *Study on task-specific QoE-oriented optimization for a video surveillance system*

1.5. Challenges

Challenges are defined by the area of improvement, which is defined by capability-oriented or result-oriented improvement. Challenges in video surveillance can be categorized into three research areas: analytics, architecture, and law and regulation. These three areas are identified to tackle separately or together.

Analytics contains intrinsic challenges which are separated as follows.

– Data pipeline: convey data by extracting valuable data.

– Knowledge modeling: creating a computer interpretable model of knowledge.

– Reasoning: translate vast knowledge and learning from data to clear explainable insights.

Architecture contains intrinsic challenges, which are separated as follows.

– Network architecture: structuring network devices and services to serve the connectivity needs of client devices.

– Computing infrastructure: composing physical and virtual resources to support the flow, storage, processing and analysis of data.

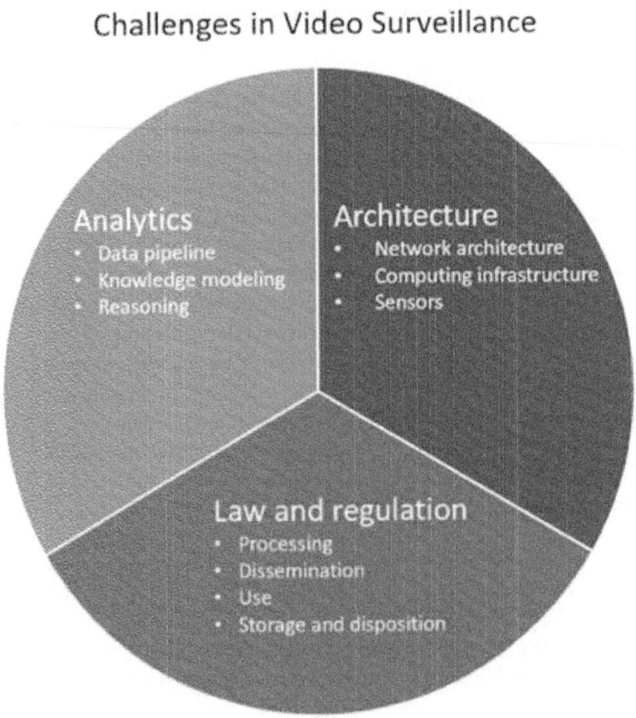

Figure 1.4. *Challenges in video surveillance*

– Sensors: integrating devices that detect and respond to some type of input from the physical environment.

Law and regulation contain intrinsic challenges which are separated as follows.

– Collecting: regulating personal data collection.

– Processing: regulating data processing.

– Dissemination: regulating data dissemination.

– Use: regulating data use.

– Storage and disposition: regulating data storage and disposition.

When referring to video surveillance:

– Analytics are subject to improvement for advanced analysis related to environment evolution, context change, spatial–temporal analysis, attention mechanism integration, explainability need, data consumption, multi-modality analysis, etc.

– Architecture is subject to improvement for advanced architecture such as UAV, wireless, mobile or satellite connectivity, for advanced resource availability as in ubiquitous environments, for advanced data capture with multi-cameras, moving cameras, 2D pan view and for advanced security solutions to protect from data deterioration and privacy.

– Law and regulation requires consideration of forbidden individual analysis, data confidentiality, availability and integrity, storage time and access, which was evoked by GDPR.

1.6. Conclusion

In the field of analytics, the study of crowd behavior is an established area of research, which is often followed by an analysis of individual behavior. Some external factors, such as the environment or traffic, should also be considered. Once a final dataset is obtained that allows for a global analysis of the urban scene, various methods can be used. However, none of these algorithms are adaptive. This leads us to ask the following question.

How can a predictive system adapt effectively to a new environment and provide support for a specific situation?

An adaptive system would allow for the analysis of an environment that evolves over time, where the context changes accordingly. It would be useful in situations where there is no prior knowledge of a new environment or when the prior knowledge is not adequate for the new environment.

In the field of architecture, many architectures have been updated to offer new capabilities through infrastructure computing and sensors embedded in their systems. Much of their work focuses on improving user-centric services regarding network architecture and computing infrastructure. However, no one has assessed the performance related to the architecture and infrastructure for the final task. This leads us to ask the following question.

How to evaluate a task-specific system that may be affected by external factors such as network and environment, as well as internal factors related to the final task, knowing it is operating in an uncontrolled environment?

External parameters such as network impact and environmental factors can deteriorate the endogenous parameters, which can lead to a specific degradation in task performance. In real-world scenarios, network and environmental factors are not controlled. Cross-area research has investigated the threats that could occur in cyberspace and impact physical space or vice versa. Even the impact of data quality on analytics has been highlighted, and how to measure it in a task-specific system. Both analytics and architecture should be studied in order to enhance performance and meet legal and regulatory requirements.

No one has studied a video surveillance system that is self-managing based on exogenous and endogenous parameters that impact overall performance, assessed according to human preference. This leads us to ask the following question.

How can a video surveillance system reduce human time and task force for management tasks?

System management can take different forms, from configuration tasks to protection tasks, through remediation tasks. For instance, manual configuration is time-consuming and adds to the workload of humans. A third-party human may introduce errors to the system configuration while not considering the final user's preference. Using automated processes in the system management plan will reduce the manufacturer's ownership and be efficient in reaction and resolution time and in task force mobilization.

2
A Development Platform for Integration and Testing

"Technology alone is not enough."
Steve Jobs

Chapter 2 presents a development platform for integration and testing. Concretely, we present a high-level view of the system, a detailed framework and then a use case for video anomaly detection.

2.1. Introduction

With the rapid evolution of technology, the amount of information available to administrations, organizations and security agents is growing exponentially. This information can come from a variety of sources, including open sources, social networks and IoT devices. Thanks to increasingly powerful and diverse terminals and accessories, such as smartphones, tablets and wearable devices, these professionals can access this information and use it to achieve their goals more effectively.

However, it is important to remember that these professionals may be working in challenging environments and situations that require innovative solutions such as artificial intelligence to provide real-time analysis and decision support. These technologies can help optimize operational efficiency and ensure mission success. To address these challenges, our system defines four research directions.

For a color version of all the figures in this chapter, see www.iste.co.uk/djeachandrane/smart.zip.

The first direction is the development of augmented situational information, which involves using AI to build real-time situational information that is relevant and contextual to the mission and situation. The second direction is intelligent interaction, which aims to simplify interactions among users, devices and equipment using AI technologies. The third direction is assisted procedure management and reporting, which involves using AI to provide effective solutions for low-value tasks such as procedural or administrative work. The final direction is operational decision support, which offers real-time decision support functionality using AI technologies that are relevant and contextual to the mission and situation.

2.2. Proposed framework – QoE-driven SA-centric DSS

As part of our research and development project, we will create a framework to improve our system. The first step will involve proposing a high-level view of the system. This will provide a simplified, abstract representation of the system's major components and their relationships, as well as the overall purpose of the system. Next, we will create a detailed framework that takes a structured approach to solving a specific problem or achieving a particular goal. This will involve developing guidelines, methodologies and processes in a comprehensive manner that is tailored to the specific problem or goal we are trying to achieve. The framework is designed to identify and resolve scientific obstacles while simultaneously ensuring that our user-centered system remains thorough and effective. We recognize the importance of addressing both application-level and scientific/technological challenges to develop a holistic solution that meets the needs of our users.

2.2.1. *High-level view of the system: reinforcement signal and QoE*

In the real-world scenario, critical service systematically involves the human factor in the decision-making loop. To benefit from it, the service must go through a user-centered design process.

User-centered design relies on an explicit understanding of users, tasks and environments and is driven by user-centered evaluation that takes into account the entire user experience. The process involves users throughout the design process – knowledge modeling and development process – knowledge assessment.

For knowledge modeling, the user preference could be tricky, narrow or even difficult to model and for that, using reinforcement signals in a trial-and-error process could afford these lacks. Additionally, AI alignment requires an assessment of user knowledge. This can be accomplished by modeling and evaluating user expectations. However, these expectations may often be unclear due to their strong connection to user preferences. Fortunately, this issue can be addressed by exploring application-oriented QoE.

2.2.1.1. *Reinforcement signal for knowledge modeling*

Reinforcement learning is a trial-and-error process where a closed-loop means adjusting the agent is possible. Without this signal, the model would not be able to learn continuously. This factor gives the model the ability to progress in a dynamic environment. This signal is known as the reward function. It indicates if the action generated by the agent is correct or not. With this information, the agent updates the policy. This signal has a real added value in a simulated environment. In a real-world situation, it can be time-consuming and labor-intensive. But as stated previously, the human factor is essential in a mission-critical operation, so we will take advantage of this and integrate it into our system.

In addition, due to the rapid change of behavior and manners in society, videos marked falsely as alarms in the DSS should be fed back to the training process to continuously improve the accuracy and generalization ability of the system. Thanks to this feedback process, the system can be considered as a context-evolving system that will update environment understanding continuously.

2.2.1.2. *QoE for knowledge assessment*

A good understanding of the environment will allow good decision-making. For this, we will take advantage of the reinforcement signal by updating the knowledge plane to include the human factor in the objective of correcting the system.

Quality of experience concerns any type of service that can be used. It represents the level of user satisfaction at the end of the data chain regarding the use of the service and includes all the objective and subjective characteristics. However, this quality of experience is measured by subjective user evaluation. Uses can be various, such as surfing the Internet, making a phone call and many more. In video surveillance, the QoE will be the metric to be gathered from the user and will be used as a reward to approve the result of the DSS.

And the use of the QoE as a metric will permit it to translate to a reward and update the knowledge plan. Here, in addition to adding context to understand the environment, it will approve and certify a good understanding of the environment in which the agent occurs in order to decrease data uncertainty and the ability to handle inconsistency. With this metric, the human factor will be at the center of our decision-support system and will have a major impact on our system and thus guarantee better decision-making.

QoE, initially used for evaluating user satisfaction with a service, is used here to bring business expertise – domain-specific knowledge in the loop and, at the same time, to validate the system understanding.

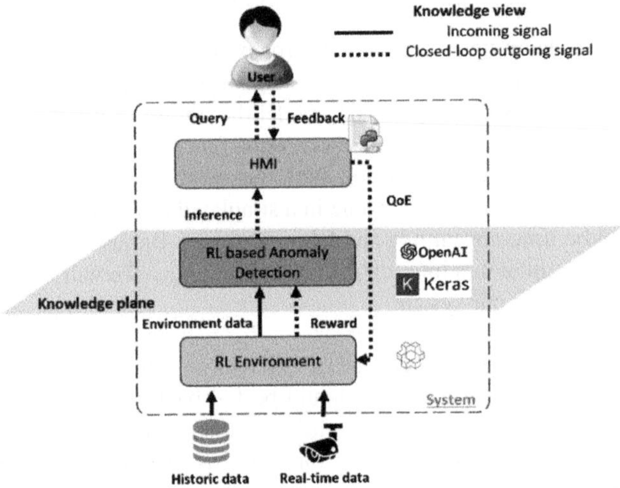

Figure 2.1. *High-level view of our proposed system regulated by QoE*

2.2.2. *Detailed system framework: SA-centric DSS*

As per our strategy, the proposed high-level view of the system in Figure 2.1 introduces a model with bidirectional communication adjusting the SA-knowledge of our DSS. The closed loop will permit us to adjust our model continuously, either to generalize the model or to introduce new abnormalities. This framework introduces a context-evolving system for mission-critical operations, including the human factor.

Our detailed system framework improves the cognition of our custom SA-centric DSS, as in Figure 2.2. It will improve levels 1, 2 and 3 of SA and assess and update knowledge. Inspired by a human-centric approach, Human-Centered AI (HCAI) is at the core of our proposal to increase the cognition level of our system. Our proposal enhances the surrounding awareness and decision support by introducing context awareness and a smart video analytics component. To improve these components' communications and interactions, a self-managed solution is explored.

This framework will permit us to answer, first, to context modeling with the context awareness component, second, to decision-making with the smart video analytics component, and third, to overall system management.

2.2.2.1. *Data enrichment for context awareness*

2.2.2.1.1. Data enrichment process

Data enrichment refers to the process of adding or enhancing collected data with relevant context obtained from additional sources. Data enrichment will produce enriched information. Enriched information is information that combines information from different sources and is represented in a format familiar to professional crisis responders as defined by Walle et al. (2016).

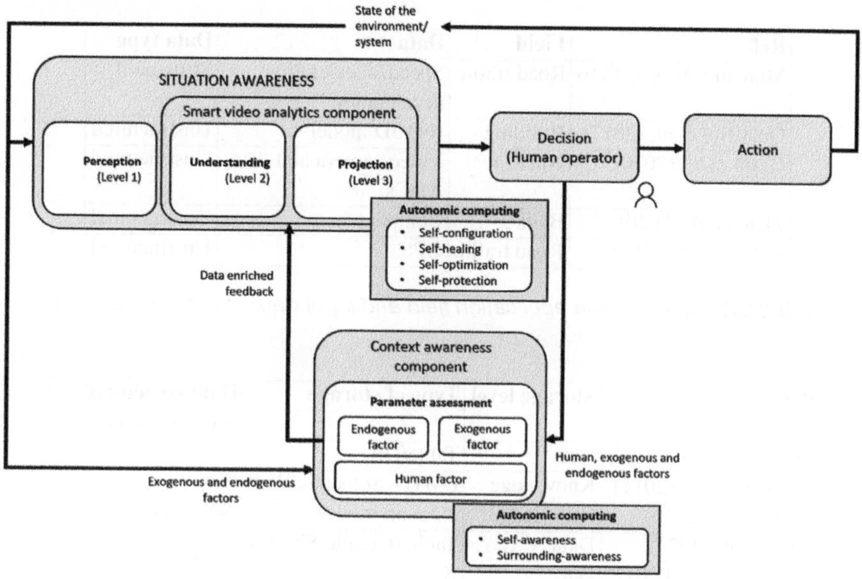

Figure 2.2. *Detailed view of our proposed SA-centric DSS framework*

This enriched information could be composed of hard and soft information. Hard information provides information on the quantitative evaluation of elements and is provided by physical sensors (camera, microphone, radar, etc.). Soft information is given by analyzing the linguistic content (observation report, text, telephone call, etc.) to qualify the elements qualitatively.

The main objective here is to align these two types of information to erase human and system drawbacks in order to provide an enhanced context awareness.

2.2.2.1.2. Review of data enrichment applications in other domains

Among methods reviewed in the literature, the data enrichment process was used in many domains and to enhance different types of data, from structured to unstructured data type, as in Table 2.1.

In Table 2.2, data is stored in proper form in a database, or simple table or even, knowledge domain representation. For those not mentioned, real-time data gathering can be conducted directly without proper storage methods, as currently utilized in their approach.

Ref.	Field	Data	Data type
Mon and Aung (2020)	Road traffic	Speed, acceleration, deceleration	Structured
Quattrini et al. (2017)	Building	IFC 3D model	Unstructured
Grahl et al. (2020)	Nuclear	Video camera and data ADCs	Unstructured
Zhao et al. (2020)	Bank	Company name	Unstructured
Rettore et al. (2020)	Road traffic	GPS	Unstructured

Table 2.1. *Study on the application field and input data in data enrichment*

Ref.	Storage level	Type of storage	Data structure/ representation
Mon and Aung (2020)	Data	Table/DB	
Quattrini et al. (2017)	Knowledge	OWL – Semantic Web	RDF
Grahl et al. (2020)	Knowledge	Domain model	Object/class
Zhao et al. (2020)	Data	Table (Google Sheets)	
Rettore et al. (2020)	Data	X	X

Table 2.2. *Study on data storage and representation in data enrichment*

In Table 2.3, data enrichment could be done in an automatic way or manually; it depends on the level of enrichment, data, information or knowledge.

In Table 2.4, it is interesting to see the final goal and the way they integrate new information into the old one to produce enriched information. Methods presented by Quattrini et al. (2017) and Grahl et al. (2020) use it as a shared platform with a particular exception for the method presented by Grahl et al. (2020) which permits data visualization in addition. Methods presented by Mon and Aung (2020), Rettore et al. (2020) and Zhao et al. (2020) use it to produce new insights from the enriched data. This enriched data can be presented in various forms, including simple new information, a multiplicative factor, or an optimization form.

This approach greatly interests us due to its ability to compute a goal by automatically fusing data and using it as an attribute in the predictive function.

Ref.	Enrichment Level	(M/A)	Method	Number of sources	Source	Data	Data type	Transformation (descriptor)
Mon and Aung (2020)	Information	A	Programming	Pre-determined/ limited	API Weather, light sensor	Meteorological, brightness, important location	Qualitative and quantitative	Quantification in reduction factor
Quattrini et al. (2017)	Knowledge	M	X	Unlimited	Multiple	Textual, sketch, comments, photos, etc.	Multi-source and heterogeneous, multi modality	X
Grahl et al. (2020)	Information/ knowledge	Semi-A	X/ Programming	Pre-determined/ limited	REST API of the logbook	Meta data and comments of various natures	Tags, video, textual, etc..	X
Zhao et al. (2020)	Model/ information	A	Machine Learning Naïve Bayesian	Pre-determined/ limited	API Yellow Page	Domain/ categories	Textual	Probabilistic model for IR
Rettore et al. (2020)	Knowledge/ model	A/X	Data fusion (Graph, PCA) and ML (SVM, KNN, RF)	Pre-determined/ limited	API Twitter, Here, Bing, Trip Advisor	Tweets, locations	Text, GPS coordinates	TF-IDF and IR

Table 2.3. *Study on the enrichment process in data enrichment*

Ref.	Data fusion	Data and/or knowledge type	Goal
Mon and Aung (2020)	Multiplication of the reduction factor to the speed limit	Speed limit	Adaptive speed limit in view of the context
Quattrini et al. (2017)	X	Contextualized information about historical heritage	Contextualized information sharing platform
Grahl et al. (2020)	Bayesian modeling with Minerva	Experimental data from nuclear fusion	Collaborative platform for sharing information and comprehensive overview of W7-X tests.
Zhao et al. (2020)	X	Industrial domain of customers	Analysis of customers according to their sector of activity to better target the financial service
Rettore et al. (2020)	Rep. Graph and ACP/X	Contextualized road traffic information	Route optimization and detection of traffic events

Table 2.4. *Study of knowledge extraction from data in data enrichment*

2.2.2.1.3. Challenges in video surveillance

Exogenous and endogenous parameters

Exogenous variables are independent and endogenous variables are dependent. Therefore, if the variable does not depend on variables within the model, it is an exogenous variable. In an economic model, an exogenous variable is a variable whose measurement is determined outside the model and imposed on the model, and an exogenous change is a change in an exogenous variable. In contrast, an endogenous variable is a variable whose measurement is determined by the model. An endogenous change is a change in an endogenous variable in response to an exogenous change imposed on the model.

For instance, as presented in our literature review, in AI, researchers assess endogenous parameters (model parameters) to improve the model; in the computer network domain, researchers assess endogenous parameters (network parameters) to improve the data dissemination and in some cases work with external factors caught as exogenous parameters to improve data dissemination over the network. In order to improve the AI model, a context-aware AI model could be developed in order to deal with exogenous (in addition to endogenous) parameters which could bring more valuable information (external factors, human expertise, domain knowledge, etc.).

Human factors

Human factors is the scientific discipline concerned with the understanding of interactions among humans and other elements of a system, and the profession that applies theory, principles, data and methods to design in order to optimize human well-being and overall system performance.

These three definitions more accurately highlight its final goal.

– Human Factors and Ergonomics Society (n.d.):

> Human factors [...] is concerned with the application of what we know about people, their abilities, characteristics, and limitations to the design of equipment they use, environments in which they function, and jobs they perform.

This definition mentions the abilities and limitations of human factors. By abilities, it could mean human expertise or reliability, for instance, and by limitation, it could mean focus or memory.

– National Research Council (1992):

> [A person] who is concerned primarily with the performance of one or more persons in a task-oriented environment interacting with equipment, other people, or both.

This definition emphasizes our ability to enhance performance in task-oriented environments. These environments can be IT-related, real-world social contexts, or a combination of both, which can be referred to as socio-technical. This task-oriented performance assessment and the environment type allow us to explore different possibilities.

– The Federal Aviation Administration (1993):

> Within the FAA, human factors entails a multidisciplinary effort to generate and compile information about human capabilities and limitations and apply that information to equipment, systems, facilities, procedures, jobs, environments, training, staffing, and personnel management for safe, comfortable, and effective human performance.

This definition emphasizes the importance of understanding human factors as valuable information for the system or environment.

The human factor has both advantages and disadvantages when it comes to completing tasks. It affects performance on specific tasks as well as the overall operating environment. This information is important to consider in order to optimize our system for human performance.

2.2.2.2. *Situational awareness for decision-making*

2.2.2.2.1. Crucial concept emerging

In decision-making, situation awareness is a key concept to understand. As stated by Mica Endsley, situation awareness drives decision-making and performance and has been recognized as a critical foundation for successful decision-making across a broad range of situations.

As defined in Endsley et al. (2003), SA is a concept introduced during World War I, evoked as a crucial skill for fighter pilots on missions. A theory of SA for dynamical systems was delivered by Endsley in 1995.

2.2.2.2.2. Definition of SA

Situational awareness is simply defined, as in Figure 2.3, by the perception of environmental elements and events in relation to time or space, understanding their significance and projecting their future status. Another way to define situational awareness is to say that it is a type of consciousness that is externally focused and adaptable, leading to an understanding of a constantly changing task environment and the ability to take effective action within that environment.

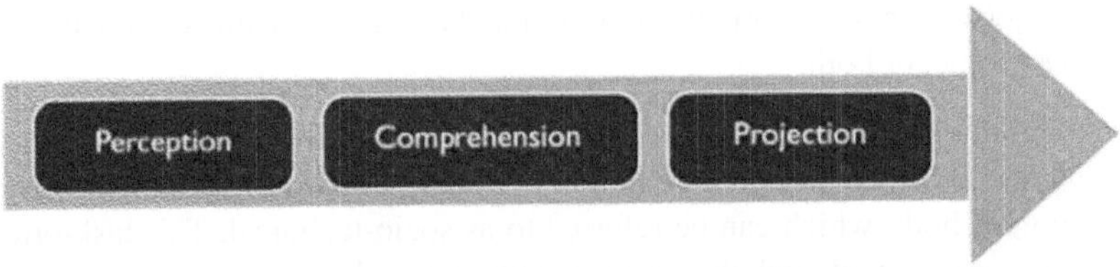

Figure 2.3. *Components of SA*

2.2.2.2.3. Theoretical SA model for a dynamic system

A theoretical model of situation awareness for dynamic decision-making was presented. And the importance of SA in decision-making in a dynamic environment has been highlighted in this study.

Figure 2.4 is the synthesized version of Endsley's model (1995).

In addition to the three components of this system – situation awareness (SA), decision-making, and performance of action – all are influenced by two main types of factors: task and environmental factors, and individual factors.

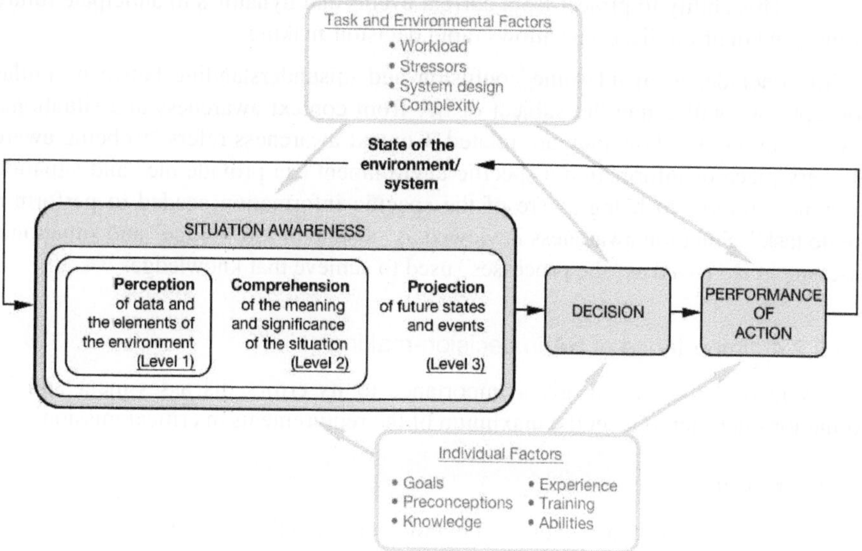

Figure 2.4. *Endsley's model of SA*

The formal definition of SA was previously defined and deeply studied in three ascending levels.

– Level 1 – Perception of the elements in the environment

- First, the perception of signals (level 1 SA) is fundamental. Without the basic perception of important information, the chances of forming an incorrect picture of the situation increase dramatically.

- Jones and Endsley (1996) found that 76% of SA errors in pilots could be related to problems in perceiving necessary information (due to system failures or deficiencies or cognitive process problems).

– Level 2 – Comprehension of the situation

- The integration of several pieces of information and a determination of their relevance in relation to the person's objectives (level 2 SA). It comes down to having a high level of understanding.

- 20% of SA errors involved SA level 2 problems (Jones and Endsley 1996).

– Level 3 – Projection of future status

- At the highest level of SA, the ability to predict future and dynamic situational events (level 3 SA) designates operators who have the highest level of situational awareness.

- This ability to project from current events and dynamics to anticipate future events (and their implications) allows rapid decision-making.

To conclude, to avoid some confusion and misunderstanding between similar concepts, we will frame the subject of SA from context awareness and situational assessment and see how they are related. Context awareness refers "to being aware of every piece of information a specific environment can provide me" and situation awareness refers "to being aware of the specific information needed to perform a single task". Situation awareness is viewed as "a state of knowledge" and situational assessment is viewed as "the processes" used to achieve that knowledge.

2.2.2.2.4. Importance of SA in decision-making

SA in decision-making gives importance to its errors, its assessment and its formations in order to meet the maximum of the requirements in critical missions:

Assessments

– SA involves the constant assessment of changes in the environment.

– SA is an important building block in the decision-making process, as it provides reliable information about reality.

Errors

– A small SA error can have dramatic consequences in critical systems.

– Differentiate between an error in the SA and an error in the decision taken.

Formation of situational awareness

– In its simplest terms, SA is derived from all our various sources of information.

– Signals may be received by visual, auditory, tactile, olfactory or gustatory receptors. Some signals may be obvious (e.g. a system alarm) and others quite subtle (e.g. a slight change in the hum of an engine) which are only registered unconsciously.

2.2.2.2.5. Review of SA application in video surveillance

Some work has been done to enhance SA in the field of video surveillance. These approaches can be listed as below:

– Multi-modal data fusion to add context to a situation (Nesen and Bhargava 2021).

– Development of an applicative-centric framework which contributes at each level of SA as "Person-Action-Locator" which analyzes video surveillance data acquired by an unnamed aerial vehicle (Geraldes et al. 2019).

– Development of a system which proceeds to do information fusion in order to make a high-level reasoning which enhances SA (Müller et al. 2017). The domain-specific knowledge was formalized to support human operators.

Few studies have been carried out in SA systems on video surveillance including the human factor. The study by Höferlin et al. (2011) involves data enrichment by users which consists of sonified trajectories of moving objects with auditory icons. They use the human factor to enrich the context in two separate phases, use and enrichment.

2.2.2.2.6. Challenges in video surveillance

The mission-critical factor will be introduced to understand the challenge in our work. A decision support system is an information system that supports organizational decision-making activities. In a control room, an enormous amount of information can be gathered. The aims of this system are to support the task force in the most relevant way. In the case of abnormal detection, the task force can operate within a short time thanks to this system. In a DSS, a mission-critical factor is any factor that is essential to mission or business operations or to an organization. Failure or disruption of mission-critical factors will result in serious impact on mission or business operations or upon an organization, and can even cause social turmoil and catastrophes. In the real world, all these analyses are carried out from the control room and are driven to the competent authorities to take actions on. With all the listed methods, we cannot approve or certify the system's reasoning and can only depend on it. We do not have any control over our model's understanding. For this reason, it is vital to have a human operator to decide instead of the system in mission-critical operations. The system is here only as a DSS. A DSS designed for mission-critical operations should have an enhanced SA. A good understanding of the environment will help decision-making. In critical systems, a bad SA can result in dramatic consequences. Hence, we want and need to update the SA knowledge to help the human operator to take a decision on a system which returns a low false rate and by reducing false positives which could transform the DSS as an inconvenient system in a real-world scenario.

2.2.2.3. *Autonomic computing for overall system management*

As presented in Götzinger et al. (2020), AC will be introduced as a pivotal concept, its initiative, its properties and its reflections to self-management and in addition ending with a review in the video surveillance field.

2.2.2.3.1. Pivotal concept emerging

The idea of autonomy in computer systems emerged in the 1990s and was influenced by biological systems (Parashar and Hariri 2004). It is projected that the amount of computing devices in use will continue to increase, along with their complexity (Horn 2001).

In 2001, IBM announced that the complexity of IT systems would be a major challenge for the industry in the future.

2.2.2.3.2. Autonomic computing initiative

To address this challenge, IBM launched the AC initiative and developed five levels of maturity: basic, managed, predictive, adaptive and autonomic (Kephart 2005). The goal of these levels was to make computer systems more autonomous and require less human intervention (Ganek and Corbi 2003).

At the lowest level, the system is managed by highly skilled staff who monitor and manually modify it based on gathered information (Huebscher and Mccann 2008). On the other hand, at the highest level, the system is fully autonomic and manages itself to fulfill high-level goals given by humans (Parashar and Hariri 2004). In simpler terms, the system is able to manage itself according to objectives provided by humans (Cámara et al. 2017).

2.2.2.3.3. Self-* properties

In addition, IBM presented the concept of self-* properties in AC (Sinreich 2006), which includes self-configuration, self-healing, self-optimization and self-protection. These properties enable the system to autonomously adjust its configuration, detect and diagnose problems, optimize resource usage and protect against malicious attacks and unintentional misuse by users.

The concept of observing and analyzing our own behavior and environment in order to adapt accordingly is known as self-awareness, and it is one of the self-* properties proposed in the autonomic computing initiative by IBM (Salehie and Tahvildari 2009). This allows a system to reflect on its own actions and make changes as needed.

For a computer system to be self-adaptive, it must be equipped with sensors to detect changes in both its internal and external environments, as well as actuators to make necessary adjustments (Parashar and Hariri 2004). The self-* properties, including self-awareness and context-awareness, form a hierarchy with self-awareness as the foundation. Essentially, a system must possess self-awareness to achieve self-adaptation or autonomy.

2.2.2.3.4. Review of AC application in video surveillance

Among all the studies carried out in AC for video surveillance, as shown in Table 2.5, among these known properties, some of those cited are self-configuring and self-adaptive for more general use. Other approaches such as self-updating, self-calibrating and self-training are offered based on their respective and specific purpose and use.

Ref.	Self-* property	Controlled system	Task
Soto et al. (2009)	Self-configuring	Camera network	Multi-object tracking for persons
Sengar and Mukhopadhyay (2016)	Self-adaptive	Object window size	Moving object area detection for person
López-López et al. (2019)	Self-updating	Predictive system	Face recognition
Tang et al. (2016)	Self-calibrating	Camera	Calibration for moving person tracking
Feng et al. (2021)	Self-training	Predictive system	Video anomaly detection from abnormal situation

Table 2.5. *Comparative study of self-* properties and tasks for autonomic computing in video surveillance*

Ref.	Self-* property	Method
Soto et al. (2009)	Active camera network which cannot cover all the areas for multitracking with a static camera network.	Kalman filter consensus to detect the same target states from different cameras, and optimization of local sensor utility functions leading to an optimal value for a global utility, i.e. keeping view of all targets at an acceptable resolution and some at high resolution.
Sengar and Mukhopadhyay (2016)	Adaptive object recognition regarding context.	Otsu (Liao et al. 2001) method to estimate the optimal threshold for noise reduction in the optical flow by estimating the discriminating criterion that maximizes the separability between the motion object and noise classes. By additional morphological operation, the adaptive window to detect the moving object area is computed.
López-López et al. (2019)	Adaptive predictive system regarding context.	Pseudo-labeling transforms a supervised classification task into an unsupervised one through an auto-labeling method that enables self-supervised learning using incremental SVM and online sequential ELM.
Tang et al. (2016)	Adaptive camera while relaxing assumption about camera parameters.	Using the estimation of distribution algorithm (EDA) to search for the local optimal solution for the camera calibration that minimizes the average reprojection error on the ground plane. The goal is to determine the camera parameters, including focal length, principal point coordinates, skew and angles.
Feng et al. (2021)	Adaptive predictive system regarding context.	Pseudo-label generation via multiple instance learning and self-guided attention in feature encoder which considered this approach as self-supervised learning method.

Table 2.6. *Comparative study on autonomic computing in video surveillance*

To go in-depth into these respective studies, their exact self-* properties are described with their methods in Table 2.6.

To conclude, all these methods could be categorized according to their objectives, update knowledge of predictive system (López-López et al. 2019; Feng et al. 2021), resource allocation of cameras, respectively, to its tracking goal as in Soto et al. (2009), initialization of the camera or recognition system (Sengar and Mukhopadhyay 2016; Tang et al. 2016).

2.3. Use case – Airbus DS SLC's target market

2.3.1. *Introduction*

The purpose of the use case is to define and evaluate artificial intelligence that will process real-time data streams (essentially videos) in order to do on-the-fly detection of situations presenting a "risk" with respect to a nominal situation. CCTV has recently been widely applied due to the security concerns raised. Despite this wide use of CCTV systems, conventional manual analysis to label anomalous events in the amount of video data gathered from surveillance of crowds and public places is time-consuming and inefficient. The chosen use case is therefore the study of urban crowd scenes due to a greater concern for security in public places, such as metro stations, airports and shopping malls. The "urban crowd scenes" use case will enable the implementation of an intelligent system that will detect anomalies using automatic learning methods. This artificial intelligence should eventually be integrated into a secure communications multimedia platform implementing the 3GPP MCS standard. As such, this artificial intelligence is evaluated alongside mission-critical systems that could contribute to the enrichment of contextualized data. This artificial intelligence is intended to help operators manage very dense and sometimes complex situations (e.g. rescue on a large-scale accident, identification of atypical situations requiring priority treatment among a multitude of data fed back simultaneously from the field, etc.). To define the Airbus SLC final use case, first, we will define the use case's inner challenge and goals before proposing an application case and, finally, the implementation case.

2.3.2. *Challenges*

Logistics for crisis management and decision support for professionals responsible for civil security (police, firefighters, crisis managers, etc.) respond to the following challenges:

– the explosion of data sources: human operational staff cannot collect and analyze all these data;

– budget restrictions and reduction of time to be allocated to administrative tasks;

– situations generating stress and the need to benefit from decision support that is not subject to emotions;

– lack of resources for major events.

2.3.3. *Purposes*

To deal with the challenges, they can be met by meeting the following purposes:

– analyze different scenes in a dynamic context and using data collected mainly but not exclusively from the critical collaboration platform: identify an incident, understand it, provide the elements to trigger the means to be mobilized (firefighter or police) and propose decision recommendations;

– pre-process data/combine data from different sources;

– detect incidents that will require the intervention of the public security service (reduction of false alerts, avoid false negatives);

– anticipate in order to avoid critical situations (e.g. accident with a truck transporting dangerous products);

– efficiently distribute relevant information to the various stakeholders by relying on group communication mechanisms and location of critical communication services;

– accompany the actors with relevant information with strong real-time and security constraints;

– eliminate multiple reports of the same incident.

2.3.4. *Application case: Airbus DS SLC's business opportunity*

In this section, we present our reflection and methodology to refine the scope of our final solution use case.

2.3.4.1. *Urban analytics*

Urban analytics draws our attention in the field of video surveillance due to its high importance previously mentioned and its conformity with the Airbus SLC's scope of application, customer needs and market.

In Zhang et al. (2020), the process of analyzing urban anomalies through data-driven logic was illustrated. To establish a connection between cyberspace and physical space, three key concepts are defined as follows.

– Urban data: cities generate spatio-temporal data through distributed sensors and gather from mobile devices.

– Urban dynamics: these pertain to the alterations in the physical state of the city, encompassing human movement such as crowds and traffic flows, as well as changes in the surrounding environment.

– Urban events: these events are the underlying causes of urban dynamics, which refer to the changes in the physical state of the city.

In order to grasp the concept of data-driven urban anomaly analysis, we begin by demonstrating the interconnections between these three concepts as they relate to real-world causality. Based on the aforementioned logic, we adhere to a comprehensive framework of data-driven urban anomaly analysis. The three components of the framework correspond to the following questions.

– What kinds of urban data are used?

– What are the general analytics methods?

– What kinds of events can be detected?

To address the above questions, we systematically reviewed relevant research from recent years. Our conclusions are presented in the "Literature Review" section, where we provide an overview of related works, specifically focusing on the data and methods used for detecting abnormal situations in urban areas. To answer this question, we will describe the degree of analysis, the detectable abnormal events and the available knowledge that we wish to deal with.

2.3.4.2. *Degree of analysis in real-world scenarios*

Inspired by Chaaraoui et al. (2012), in Figure 2.5, human-related analysis tasks are classified into motion, action, activity or behavior levels regarding the degree of semantics and the amount of time involved in the analysis. Therefore, it shows that both the time frame taken into account and the degree of semantics (DoS) involved in the recognition and classification process grow as we reach a higher level of the pyramid.

Not only time frame and semantic degree grow at higher levels of this hierarchy, but also complexity and computational cost lead to heavy and slow recognition systems, as each level requires most of the previous level tasks to be done too. According to this work, we can conclude that behavior analysis stands as the most challenging task in video surveillance.

Abnormal behavior recognition from videos is one of the long-standing problems in computer vision and machine learning with wide-ranging applications in surveillance. Millions of surveillance cameras are being deployed in public and private places requiring intelligent video monitoring. While video content analysis (VCA) has a very large collection of applications in a surveillance environment, one such focused area of research is anomalous behavior recognition. Abnormal behavior recognition deals with identifying the patterns and events that vary from the normal stream.

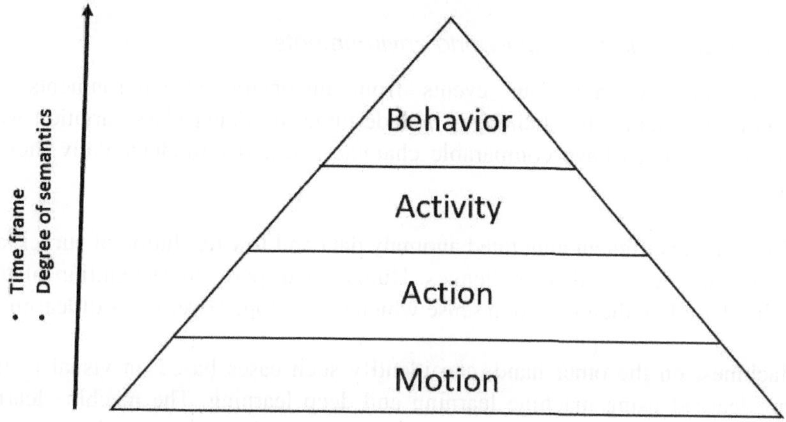

Figure 2.5. *Human behavior analysis tasks – classification*

2.3.4.3. *Real-world anomalous events*

Anomalies contain a huge range of activities that can go from abuse to fighting and road accidents to snatching. These events take place in different surroundings (such as public buildings and roads, etc.) and during different times of the day and weather. Usually, anomalous activities occur infrequently as compared to normal activities. Therefore, the development of an automated algorithm for anomalous video activity recognition is an urgent concern, as manually handling a high volume of video content is costly and error-prone due to human limitations.

From our study, individual behaviors that disrupt public order are a particular focus of our research into anomaly detection in urban surveillance videos. This has become a new challenge since the success of deep learning for analyzing crowd behavior by introducing more complex anomalies instead of unexpected crowd clustering or dissolution. It became a hot topic for the public safety community initially because of “asymmetric threats” (Pavlidis et al. 2001). Lately, a lot of research has been carried out on video surveillance footage analysis. These studies show that a particular interest is portrayed in behavior analysis (Popoola and Kejun Wang 2012) in urban scene analysis. This individual behavior anomaly can be analyzed when it is caused by humans, or when it may have an impact on humans due to non-human behavior, such as traffic (surrounding traffic information, car damage, sudden car stop or even accident) or the environment (fire, explosion, detonation light).

Detecting abnormal situations from crowds to individuals through environment and traffic analysis in urban scenes is a concern of video surveillance, as it can help to identify a potential high value situation and prevent it from occurring. To avoid confusion, we switch the term behavior to situation to match with crowd, individual, traffic and environment.

2.3.4.4. *Knowledge from real-world environments*

Recognition of anomalous events from unconstrained environments is a challenging problem. The challenges include huge intra/inter-class variation where two or more classes have comparable characteristics, but fundamentally they are different.

Moreover, insufficient annotated anomaly data and low resolution of surveillance videos are among the other challenges. Humans can perform recognition of these anomalies based on their common sense which is developed over years of learning.

Machines, on the other hand, can identify such cases based on visual features that are learned using machine learning and deep learning. The machine learning algorithms can perform better if provided with unique features having better discriminatory power between different classes.

2.3.4.5. *Definition of the use case*

Our use case was done regarding the degree of analysis in real-world scenarios, real-world anomalous events and available knowledge from real-world environments. Regarding these aspects, our use case will consist of detecting general abnormal situations from video footage.

As mentioned in our literature review, in 2018, a large-scale CCTV dataset called the UCF-Crime dataset (Sultani et al. 2018) was introduced that is composed of 13 anomalous events captured from surveillance cameras. This dataset has been chosen because it contains high value situations, varied events, complex data (irregular), weak labels and unbalanced events.

Motivated by the aforementioned limitation, the objective of our study is to make proposals without avoiding the sub-challenges of data-based analysis that are part of the problematics of real scenarios.

Moreover, the use of this kind of technology risks undermining privacy and civil liberties, as well as prejudice and discrimination. It is therefore important to strike a balance between the benefits of this technology and the need to protect privacy and civil liberties, and to ensure that it is used fairly and impartially. This concern has its role to play in a society where privacy and civil liberties are at the heart of the thinking of solution providers such as Airbus DS SLC in Europe where the rules are dictated by the GDPR.

2.3.5. *Target system: Airbus DS SLC's flagship product*

The target system is described in Airbus DS SLC (n.d.) by presenting first its use, its functioning and finally its benefits.

2.3.5.1. *What is the Airbus DS SLC's flagship product?*

The flagship product provides a secure way for groups to collaborate, bringing professional communication to smart devices. It allows radio and smart device users to join the same groups, giving smart device users access to voice, data, video and location services in a professional and reliable manner. From simple push-to-talk to a comprehensive group collaboration solution, this flagship product offers a range of communication methods and tools to manage users and groups effectively. It caters to the needs of public safety, transport, utility and industry users, enabling easy and secure access to staff and data across different devices and technologies. With this product's mission-critical services, professional smartphone users can easily share voice, video, multimedia, files and location information with a defined group of people at the push of a button.

2.3.5.2. *What is it for?*

The flagship product offers a comprehensive range of communication tools to meet the diverse needs of public safety organizations, transport, utility, industry and corporate sectors. With integrated applications such as voice services, instant messaging, video communication, location mapping and emergency calls, this flagship product ensures reliable and professional communication. It enables easy and secure access to staff and data across different devices and technologies, providing a strong group collaboration solution. By adopting this flagship product, smart devices and applications are integrated seamlessly into operational procedures, as well as with the existing communication infrastructure.

Figure 2.6. *Application use in control room. Source: https://www.secureland communications.com/agnet-tetrapol-p25-dmr-analogue-connectivity)*

2.3.5.3. *How does it work?*

With the professional group communication feature, connecting with the team is easy using just your smartphone. Just press a button to talk, send multimedia messages or even live video to a group. Sending or receiving real-time video from the field can be done, providing a better picture of the situation. This way, it is possible to stay ahead of the curve instead of just reacting to what has already happened. Additionally, sharing a tactical map of an operation can provide a comprehensive overview and help to plan and execute the correct response. And if you need to communicate with just one person, it is possible to make an individual voice or video call and send additional information with an individual message.

2.4. Conclusion

This chapter introduces a high-level view of the system to provide a simplified representation of the system, and a detailed, comprehensive framework is proposed to specify guidelines, methodologies and processes that are tailored to the problem. Both are designed to simplify the integration and testing processes for development platforms. This framework is designed to efficiently tackle problems in scientific and technical domains, without compromising on the ease of use for the end user.

At the high level, we use reinforcement learning to model knowledge and assess it with Quality of Experience (QoE). Reinforcement learning is a machine learning technique that trains an agent to make decisions in a dynamic environment by maximizing rewards. Our framework uses it to model knowledge and enable the system to learn from experience. QoE, on the other hand, measures the overall experience a user has with a system and is used to assess the effectiveness of the knowledge modeled by the reinforcement learning agent.

At a detailed level, the framework incorporates data enrichment for context awareness, situational awareness for effective decision-making and autonomic computing for efficient system management. Data enrichment involves enhancing data with additional information to provide context and facilitate analysis. Situational awareness is the ability of the system to perceive and understand its environment and use that understanding to make informed decisions. Autonomic computing is a self-managing computing model that enables a system to adapt to changing conditions without human intervention.

Furthermore, this chapter details a specific use case for the company's business venture, which involves identifying abnormal situations through video surveillance footage. Our approach will be informed by the flagship product, a collaborative platform on which transits crucial mission-related information. By leveraging this feature, we can make informed choices and optimize the system's performance to accurately detect abnormal situations.

3

A Multi-Criteria Enriched Corrective Signal with Endogenous, Exogenous and Human Factors

"An ounce of algebra is worth of a ton of verbal argument."
John B Haldane

Chapter 3 describes a novel multi-criteria enriched corrective signal. The purpose is to provide information about surrounding awareness in real-world scenarios regarding endogenous, exogenous and human factors.

3.1. Context

In critical situations, decision-making should be based on accurate context awareness to prevent disruptions or critical outcomes. As previously mentioned in Chapter 2, decision-making has limitations for both humans and systems. Therefore, the goal is to combine the strengths of both to provide a more robust awareness of the context. In decision support systems, humans are mandatory at the decision stage.

To make the most suitable decision, it is essential for a system to have context awareness. This context can be enhanced in various ways, ranging from raw information to valuable information. Data is an elementary description of reality, such as an observation or a measurement. Information, on the contrary, is given data plus meaning, while knowledge is based on a collective repository and shared.

From an individual's perspective, data, information, and knowledge are defined as articulated in Belmondo (2006), which is inspired by the work of Grundstein and Rosenthal-Sabroux (2001).

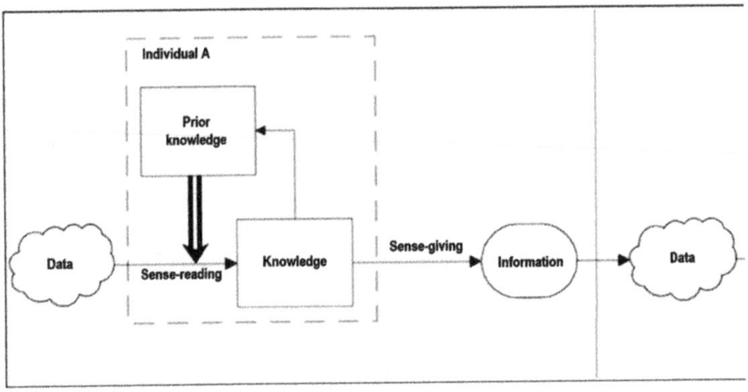

Figure 3.1. *Relationships between data, knowledge and information*

Sense reading is when knowledge is requested to understand the sense of the gathered data. Sense giving is when knowledge and data are coupled to produce new insight called information. Data is primordial because it allows for analysis of the environment, information is primordial to understand the context, and knowledge is key to giving sense to data and reading it. A knowledge base is crucial for a good understanding of the environment and context.

In computer science, a knowledge base gathers knowledge specific to a given specialized field in a form usable by a computer. The process of building knowledge begins with the collection of data and information. The chain of data should be accurately defined in systems, and the chain of information should be accurately defined for humans. The control over the chain of data provides reliability over the system, and the control over the chain of decisions provides trustworthiness in decision support for humans. Building a knowledge base can be done manually or automatically in the computer science and AI domain.

In human cognition, learning can be formal or informal, explicitly designated for the former and unintentional for the latter. In expert systems, translating informal knowledge to formal terms defined by logical notation can be challenging, especially when the knowledge is uncertain. In deep learning, it can deal with a high number of variables but will only be as good as the unbiased data available in sufficient quantity.

The power of machine learning (ML) is that it can discover patterns, formal or informal knowledge, without being directly programmed. However, its limitation is the data provided to this kind of system. Data is the raw version of itself, and information is the data with its description, whereas knowledge is the data descriptor. In ML, information is the prior knowledge where it and data are necessary to orient the learning. Prior-knowledge agnostic ML is seen as unsupervised methods which have their utility for other cases, such as pattern discovery or data compression, but not in this case. In ML, the dataset consists of data X and prior-knowledge Y. Prior knowledge is always attached to data. When it comes to the dataset, it has limitations and problems in ML, which is the training dataset problem.

Just like everything, these systems are not perfect. The training data issue is one of the most significant challenges that gets in the way of AI projects within organizations. With training data, more often than not, one of the following happens.

– The data is filled with biases or errors.

– There is not enough data to train systems.

– Tools are not available to label the data.

– The data are present but are not in a usable form.

To get the context of the adding problem, we will formulate the dataset as a couple of [X,Y], where X is the independent variable and Y is the dependent variable, which depends on changes in the independent variable.

To add context, we have the following scenarios.

– New X and new Y: more data and more prior knowledge, which drop into the data availability and label cost problem.

– New Y with existing X:

- human expertise for reliable prior knowledge (human in the loop);

- data analytics (data mining, knowledge extraction, etc.) for value-added prior knowledge.

In this section, we need to build a corrective signal and determine which contextual information is crucial for making decisions, as well as how to reliably capture it.

3.2. Problem statement

To enhance the contextual understanding of our predictive system, we need to consider both human preferences and all the factors that affect the system, whether they are endogenous, exogenous or human. Therefore, the signal must be designed in a way that takes into account all the relevant criteria. This led to a multi-criteria decision-making problem formulated as follows in the criteria space:

$$\begin{aligned} &\text{"max" } \mathbf{q} \\ &\text{subject to} \\ &\quad \mathbf{q} \in \mathbf{Q} \end{aligned}$$

where $\mathbf{q}$ is the vector of k criterion functions (objective functions) and $\mathbf{Q}$ is the solution set, $\mathbf{Q} \subseteq \mathbf{R}^k$.

Q could be defined explicitly by a set of alternatives or implicitly by a set of constraints.

The criteria values will be consequences of the decisions we make. Hence, we can define a corresponding problem in the decision space when using a predictive system. The multi-criteria decision-making problem is formulated as follows in the decision space:

$$\max q = f(x) = f(x_1, \ldots, x_n)$$

subject to

$$q \in Q = \{f(x) : x \in X, X \subseteq \mathbb{R}^n\}$$

where $\boldsymbol{X}$ is the solution set, and $\boldsymbol{x}$ is the decision variable vector of size n.

In order to address the Multiple Criteria Decision Analysis (MCDA) problem and integrate preferences, the problem can be formulated as a multi-attribute utility function. This function helps to identify the most preferred alternative or rank order the alternatives. The utility functions are expressed as a function of a bundle of goods or services, denoted as $U(X_1, X_2, X_3, X_n)$.

In this chapter so far, the problem was formulated as an optimization function modeling problem in the spatial domain. However, to increase situational awareness, the utility could be defined in the temporal domain. In the temporal domain, a time-utility function specifies the application-specific utility that an action produces within a specific time frame. For a better understanding of this concept, see Chapter 4, where it will be integrated through an inner time-evolving process into the predictive system.

3.3. Proposals

In order to understand consumer preferences for a set of goods and services, we measure them using a utility function. This function is constructed through a process called data enrichment, which involves adding or enhancing collected data with relevant context obtained from external sources and filling in missing or incomplete data. The enrichment process can be done with both objective and subjective information, where subjective information is based on personal interpretation, while objective information is based on factual data.

To guide future choices and improve context awareness, we looked at how existing video surveillance methods handle this issue. We found that approaches in spatial space consist of analyzing scenes at both a macro and micro level. The micro level involves analyzing individual behavior using object-level detection, while the macro level involves analyzing the global-level spatial relationship of a group of entities.

Our approach will use a mixture of these two types of analysis to cover a maximum of information and analyze individuals' behavior.

Another challenge we face is that SA was initially created as a social system for humans, but now we are in a technical system. To study the social impact that could operate on the technical system, we need to formulate the problem as a sociotechnical system (Militello et al. 2014). This requires AI–human collaboration to afford this lack in this kind of system.

To model the function that satisfies human preference, we investigate end-user subjective information called user experience. User experience describes how much somebody enjoyed or disliked using something, including the factors that contributed to their overall perception of it.

The quality of user experience can be measured using a metric called the Quality of Experience (QoE) (Mellouk et al. 2013; Mushtaq and Mellouk 2017). In the field of computer networks, QoE tools are used in conjunction with Quality of Service tools. This helps to identify any degradation in service from a user's perspective, which can then trigger further analysis and remedial actions.

Traditional QoE typically takes an end-to-end approach, either by evaluating the service through a group of users or by simulating user behavior using specialized tools. It quantifies satisfaction by assessing the environment, which can be either controlled or uncontrolled.

A controlled environment has control over parameters, while an uncontrolled environment does not. Endogenous parameters refer to any parameters related to a model, such as neurons, layers, weights, bias, learning rate, discounting factor, exploration function, policy and rewards. Exogenous parameters are those that come from external sources and are not directly related to the internal parameters of the smart component, such as signal loss, data masking, latency, packet loss, data poisoning, signal strength and signal noise.

In a controlled environment, both endogenous and exogenous parameters can be assessed, but in an uncontrolled environment it is more difficult to do so. Exogenous parameters have an impact on endogenous parameters. QoE subjective assessment allows for modeling of the quality of endogenous and exogenous parameters in a controlled or uncontrolled environment from an end-user perspective. This metric means dealing with a partially observable environment is possible.

Our intelligent component can model all network parameters and external parameters as exogenous parameters. The environment can be modeled using computer network parameters (Marfievici et al. 2013; Jeftenić et al. 2020) such as connectivity and performance. Our approach based on QoE is also compliant with any network architecture, while certain QoS metrics are relevant for one architecture but less so for another. For example, signal strength might be relevant in an aerial surveillance scenario but less so in CCTV surveillance. However, using an end-to-end user perception method will allow QoE to be assessed on any architecture.

Our approach to building the human utility function involves using a task-specific QoE metric, adapted to our mission of analyzing video content in an ever-changing environment. We propose to build the metric by objectively assessing the quality of data and subjectively assessing the quality of knowledge. Both assessments will help satisfy our final solution, as ML learns from knowledge, and data drives it. In an evolving context, our system should be aware of data and knowledge quality over time. Data quality will help model exogenous parameters, and knowledge quality will help model endogenous parameters. However, these assessments will not be satisfactory for the final use and will be inefficient if the video flow is corrupted or if the camera field of view is obstructed.

To ensure data quality, we will use established QoE metrics recommended by the computer network community. These metrics evaluate the quality of the signal and perceived image while taking into account end-user preferences. If the image is difficult to read or unsuitable for the mission, the quality level will decrease. In addition, to model knowledge quality, we will use Quality of Prediction (QoP) metrics as recommended by the AI community. These metrics detect anomalies perceived by the end-user, and better detection will result in improved quality. Moreover, we propose two additional augmented QoE metrics to satisfy video surveillance final users, one for assessing video integrity and another for detecting camera tampering.

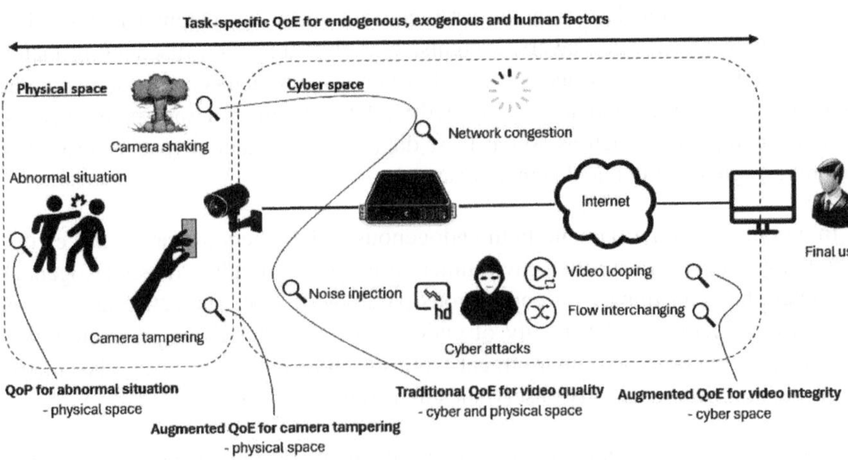

Figure 3.2. *End-to-end smart video surveillance assessment*

The composition of these metrics will create the task-specific QoE for our DSS in smart CCTV. Task-specific QoE evaluates, in the most accurate and reliable way possible, the perception of a user in a business/mission goal in an uncontrolled environment.

The following section will present our proposals:

– QoP modeling for endogenous parameters;

– task-specific QoE including endogenous, exogenous and human factors.

3.3.1. *QoP for endogenous factor assessment*

In general, ML models are trained using ground truth data available in the dataset. However, the problem is that the model can become outdated when the context evolves according to the environment. This poses a challenge in real-world scenarios where gathering ground truth data is more complicated and concerning.

To address this issue, QoP is built through an active learning process that requests signals from human operators only when uncertainty increases from one decision to another. This approach allows for an evolving ground truth and makes it possible to deal with potential context evolution.

Metadata		Agent memory		Segment-level Ground truth		QoP
Video	Segment	Reward	Action	State	Label	Score
1	1	1	0	4096 features	0	
1	2	0	1	4096 features	0	
1	3	1	1	4096 features	1	
-	-	-	-	-	-	93,75
-	-	-	-	-	-	(unseen rewards = 1)
-	-	-	-	-	-	
1	30	0	1	4096 features	0	
1	31	1	1	4096 features	1	
1	32	1	1	4096 features	1	

Figure 3.3. *Quality of Prediction with an evolving ground truth built on human feedback. For a color version of this figure, see www.iste.co.uk/djeachandrane/smart.zip*

Ground truth	Reward	Action
Abnormal	1	1
Abnormal	0	0
Normal	0	1
Normal	1	0

Table 3.1. *Anomaly detection ground truth in reinforcement learning*

As shown in Figure 3.3, from the QoP, the ground truth can be found and understood by the system. To translate from QoP to ground truth, the translation table in Figure 3.1 is followed.

Three approaches, as shown in Figure 3.4, exist in processing the reward predictor, namely direct user request, auto-labeler and user data analysis.

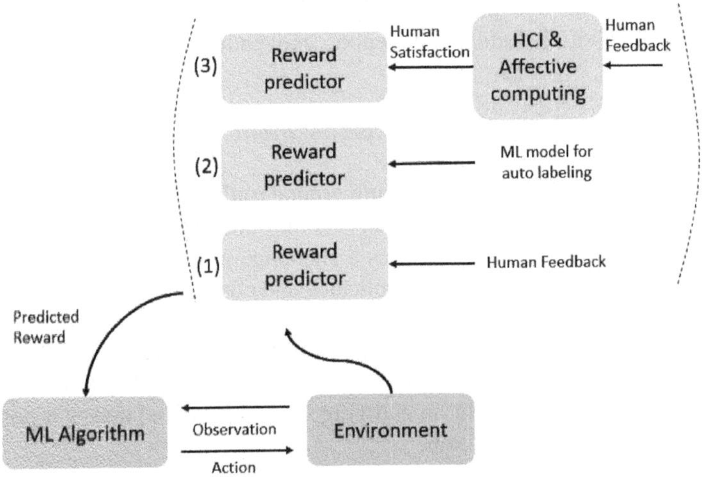

Figure 3.4. *Three approaches for ground truth gathering. For a color version of this figure, see www.iste.co.uk/djeachandrane/smart.zip*

For the first approach based on Human–Computer Interaction (HCI), a user-keyboard request is a simple way of gathering satisfaction from users, which can be directly caught by a reward predictor without passing through a conversion of user satisfaction to a system reward. Although this approach can be time-consuming, it brings reliability. In the code and the output, the user request is a straightforward method that allows the modeling of the reward in the easiest way possible. The system was simply programmed to obtain a reward of 0 and 1, where 0 represents unsatisfied and 1 represents satisfied.

For the second approach, an attempt is made to develop an auto-labeler, which is a modern way of gathering satisfaction from ML models. It can be caught directly as a reward predictor without passing through a conversion of user satisfaction to a system reward. Although this approach can be less reliable, it reduces the labor-intensive and time-consuming process. To create an auto-labeler, we will develop an ML model that will learn the full sequence of video and provide the sequence-level label, i.e. sequence learning. For this, models that catch spatial and temporal patterns should be used. In our approach, we will experiment with four models: Hidden Markov

Model (HMM), Gated Recurrent Unit (GRU), Long Short-Term Memory (LSTM) and Transformers. To conclude these experiments, these results are not significant and/or not representative of the model's capabilities and performance. The idea here was to develop a different sequence learner to provide a Proof of Concept (PoC) as an automatic labeler on the UCF-Crime dataset (Sultani et al. 2018). The best result obtained came from the transformer encoder, but further research should be carried out properly with an experimental plan. The order of implementation was in the complexity order of the model, starting from a statistical model to a deep learning model with the fewest parameters known as GRU towards a more complex RNN known as LSTM, and towards a huge model, which is the transformer encoder. The algorithms were implemented in the order in which they were introduced.

This third approach has caught our attention and, in this part, we propose an approach that combines the advantages of the two previous methods, i.e. user request and automated labeling. This approach consists of developing an automatic method to catch user satisfaction called affective computing. Two systems will be developed for two different scenarios.

– Scenario 1: gathering the QoE from the control room.

– Scenario 2: gathering the QoE from the field by first responders through the Agent MCX solution.

In order to develop our two systems, we have leveraged pre-existing libraries and APIs. To provide greater clarity on their inner workings, we will offer a concise breakdown of the methodologies we used. Each system was developed with a distinct use case in mind, allowing for tailored approaches. Our primary objective was to enhance the fidelity of capturing user reactions and to increase the reliability of both systems. To accomplish this, we proposed two different methods. The first approach involves eliminating all extraneous actions that are not linked to the final task in the sequence. This approach yields a more trustworthy user satisfaction rating for the system. The second approach entails merging data at the decision-making level. We will explore two modalities to bolster the system's dependability. By adopting these two approaches, we aim to achieve a higher level of precision and reliability in our systems. The first approach will be described in the following section.

3.3.1.1. *Human–computer interaction and vision-based affective computing*

3.3.1.1.1. Proposed method

Our proposed system is composed of four components, as in Figure 3.5.

The workflow involves several steps. First, it detects the mouse activity and confirms that it is happening on the monitoring system. Second, it detects the face as a pre-filter. Third, it tracks the eyes and confirms that they are looking at the monitor. Fourth, it detects the face again as a pre-filter. Finally, it detects the user's satisfaction through the emotions related to their task. The emotion recognizer and eye tracker both need only one face detector, even though it was developed separately for each solution due to the API limitation.

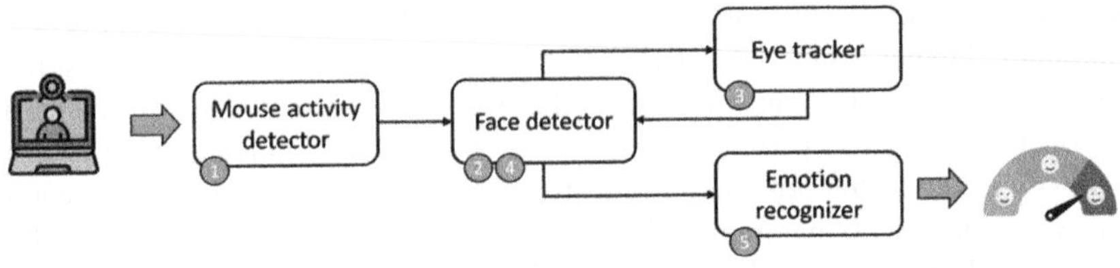

Figure 3.5. *HCI and vision-based QoE computing system workflow*

To develop this overall system, a mouse tracker, face detector, eye tracker and vision-based emotion recognizer will be developed.

Mouse tracking

Pynput library (Palmér 2021) is used in our system as the first step. This library allows for handling the mouse through a mouse listener, which is a thread where all callbacks will be invoked from the thread.

This approach was developed in order to catch activity in the monitoring system in the control room directly from the computer station.

Eye tracking

Antoine Lamé's gaze tracking library (Lamé 2019) is used in our system as the second step. This library permits the webcam-based eye tracking solution itself based on the DLIB library (King 2009), which is an ML library. Eye tracking works in two phases: first, it detects the face bounding box, and second, it detects eye landmarks.

The face detection is done with the function face_detector of the DLIB library by using a Histogram of Oriented Gradient (HoG) and linear Support Vector Machine (SVM). The eye landmark was done with the function shape_predictor of the DLIB library by using an ensemble of regression trees as done by Kazemi and Sullivan (2014).

This approach was developed in order to catch activity in the monitoring screen in control rooms directly from the computer station. It permits the link between the computer on which the user interacts and the video surveillance monitor on which the user has attention.

Vision-based affective computing

The DeepFace library (Serengil and Ozpinar 2021) is used in our system as the third step. This library permits the analysis of facial attributes such as age, gender, emotions and race. And the CV2 library (Itseez 2000) for the face detector, which is a library designed to solve computer vision problems.

It first uses a face recognizer with an Adaboost classifier and then uses CNN-FNN for emotion classification. For our application, we modified the output to provide positive rewards directly for neutral reactions, along with negative rewards for sad, disgusted, fearful, happy, angry and surprised reactions.

This approach was developed in order to catch user emotion and satisfaction in the monitoring screen in control rooms directly from the computer station. It permits the linking of the video surveillance monitor on which the user has attention and the emotion gathered from this monitoring activity.

3.3.1.1.2. Experiments

In the proposed system, the output consists of obtaining mouse tracking logs to detect any movement in computer usage, as in Figure 3.6, and then proceeding to an eye tracking of the user through the webcam, as in Figure 3.7, to detect its attention on the correct monitor and finally, to catch the user's final emotion by a vision-based affective computing method as in Figure 3.8.

```
2022-04-05 17:35:07,635: Mouse moved to (1684, 1117)
2022-04-05 17:35:07,666: Mouse moved to (1701, 1116)
2022-04-05 17:35:07,693: Mouse moved to (1723, 1114)
2022-04-05 17:35:07,723: Mouse moved to (1817, 1085)
2022-04-05 17:35:07,756: Mouse moved to (1847, 1067)
2022-04-05 17:35:07,785: Mouse moved to (1857, 1060)
2022-04-05 17:35:07,818: Mouse moved to (1864, 1060)
2022-04-05 17:35:07,845: Mouse moved to (1883, 1059)
2022-04-05 17:35:07,883: Mouse moved to (1897, 1056)
2022-04-05 17:35:07,913: Mouse moved to (1905, 1051)
2022-04-05 17:35:07,938: Mouse moved to (1911, 1049)
2022-04-05 17:35:07,973: Mouse moved to (1914, 1049)
2022-04-05 17:35:08,273: Mouse moved to (1890, 1049)
2022-04-05 17:35:08,304: Mouse moved to (1822, 1049)
2022-04-05 17:35:08,336: Mouse moved to (1737, 1051)
2022-04-05 17:35:08,366: Mouse moved to (1668, 1060)
```

Figure 3.6. *Mouse tracker system output*

Figure 3.7. *Eye tracker system output (source: webcam from computer). For a color version of this figure, see www.iste.co.uk/djeachandrane/smart.zip*

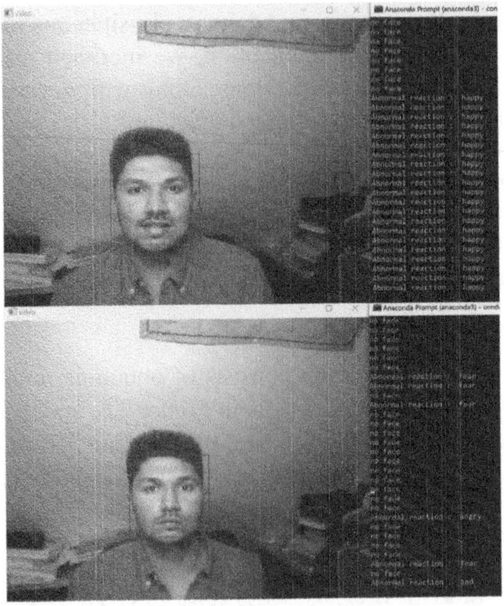

Figure 3.8. *Vision-based affective computing system output for satisfied and unsatisfied emotions (source: webcam from computer). For a color version of this figure, see www.iste.co.uk/djeachandrane/smart.zip*

3.3.2. *Task-specific QoE for endogenous, exogenous and human factors*

The QoP should provide information about the quality of the decision support. The ultimate goal is to improve this method, which can be achieved by gathering all the conditions that humans unconsciously consider in their chain of information from the perceived content to their cognitive mechanism. This can be identified through human preferences. It is important to note that the QoP metric can be easily misled if the quality of the data being used decreases, the content is not visible, or if it is evaluated on corrupted data. To address this, a metric that covers the entire chain of information should be developed. From a system perspective, this chain of information falls at the level of data and should be aligned with QoE to enable the evaluation of the decision support system from end to end, while the chain of data remains uncontrolled.

In this section, we propose to develop a task-specific QoE to model overall system performance, taking into account endogenous, exogenous and human factors. To achieve this, we will first select a traditional QoE metric that models exogenous factors in the field of video data dissemination. Then, we will create two new assessments related to the final task, called augmented QoE. Finally, we will propose our task-specific QoE metric.

3.3.2.1. *Traditional QoE for video quality assessment*

To create the traditional QoE, many automated assessment methods exist, among them, Video Quality Assessment (VQA) and Image Quality Assessment (IQA) will interest us for our application case, video processing. This QoE can be impacted by both physical and cyber spaces. It can be triggered if the data quality is affected due to actions happening in the physical space, such as explosion-type vandalism that may shake the camera (voluntary) or a natural catastrophe (involuntary). In cyberspace, QoE can be affected by network traffic congestion (involuntary) except for in Denial of Service cases or data noise injection attacks (voluntary).

Image and VQA deal with quantifying the quality of an image or video signal as seen by a human observer using an objective measure.

Video/image quality assessment (V/IQA) is crucial for the optimization of several video processing applications. And among these methods, many categories exist, full reference (FR) compares distorted video/images with the original one, reduced reference (RR) compares the reduced information from the original video/image with the corresponding information from the distorted one, and no-reference (NR) does not require the original video/image and searches for artifacts in the pixel domain of this last.

In our video processing, we did not have full or reduced information about its quality before capturing in a control room, so we cannot compare it with source information as a reference.

3.3.2.1.1. Selected method

To create our traditional QoE, we prefer to use the VQA method over the image one because it can relate to the time domain (Wang et al. 2020; Galteri et al. 2020) more and provide more accurate information. But considering IQA remains an option. NR-VQA is interesting if distortion appears after we have chosen a solid source of information; it will compute the distortion. NR-IQA will be interesting to compare in general the quality of the video without previous frame relation.

We selected the VIIDEO algorithm (Mittal et al. 2016; Ghadiyaram et al. 2017) on the basis of NR-VQA results obtained on the LIVE VQA dataset for subdataset resemblance (pedestrian area, park run, rush hour, for instance), performance cited and its ease in implementation (Sinno and Bovik 2018).

The VIIDEO algorithm performs much better than the legacy FR quality measure Mean Square Error (MSE) on the LIVE VQA database and delivers performance comparable with a leading human judgment trained blind VQA model. This NR-VQA approach is based on intrinsic statistical regularities observed in nature and when deviations from these regularities occur, it alters their visual impression. Quantifying measures of regularity (or lack thereof) within a statistical model of natural video allows for an assessment of the quality of the video. Even though it is automatically computed with no reference, this metric satisfies comparable results with human judgments.

3.3.2.1.2. Experiments

In Figure 3.9, to link knowledge quality impacted by data quality, we assessed UCF-Crime data quality with the VIIDEO algorithm, an NR-VQA method, on the testing set. Data quality shown here is mainly approximately 0.8 which is a top-of-the-basket value. Some of them are really bad at 0.2, and some are in particularly good quality at over 1.

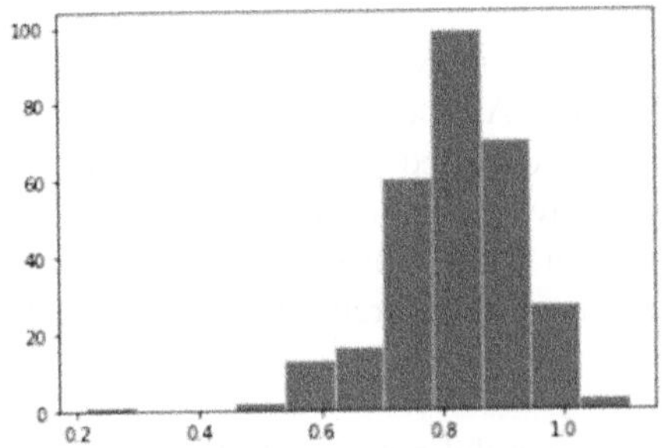

Figure 3.9. *Traditional quality of experience with no reference video quality assessment metric called VIIDEO on testing data*

In Figure 3.10, we display on the left, a frame from a video with the lowest VIIDEO score and on the right, a frame from a video with the highest VIIDEO score. "Normal_videos_019_x264.mp4" shows footage from a CCTV camera installed indoors with low impact from the exterior environment and remains in good constant quality all along the video. "Explosion021_x264.mp4" shows footage from a CCTV camera installed outdoors in a gas station with a higher environmental impact which is real as the quality was impaired by the explosion of the gas station and created a camera shake and near-blind blurry vision. As mentioned, this dataset already contains uncontrolled noise, so a study with simulated noise would be biased by already existing noises. Nevertheless, this real-world dataset shows the existence of noise in real-world scenarios that is quantifiable with the VQA VIIDEO method.

In Figure 3.11, the scores are provided alongside the video "Explosion021_x264.mp4". This video contains a replayed explosion with a short delay between the first and second. As this algorithm uses references between different frames and analyzes frame statistics between frames, the choice of the number of frames is important; it will track irregularities in a time window. In the time window, if the first frames already have noise, it will not detect a decrease of quality because it will consider it as constant quality, hence not irregular. For example, in the figure, the first explosion has more normal situation time before the explosion occurs, so the score is lower than when the second appears because it has less normal situation time.

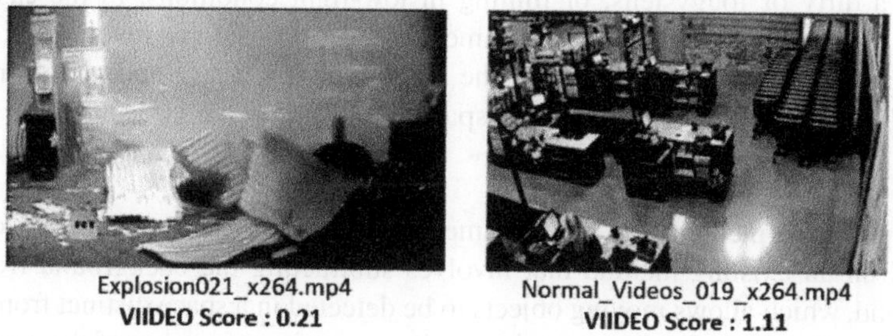

Figure 3.10. *Lowest and highest VIIDEO score as traditional QoE indicator. For a color version of this figure, see www.iste.co.uk/djeachandrane/smart.zip*

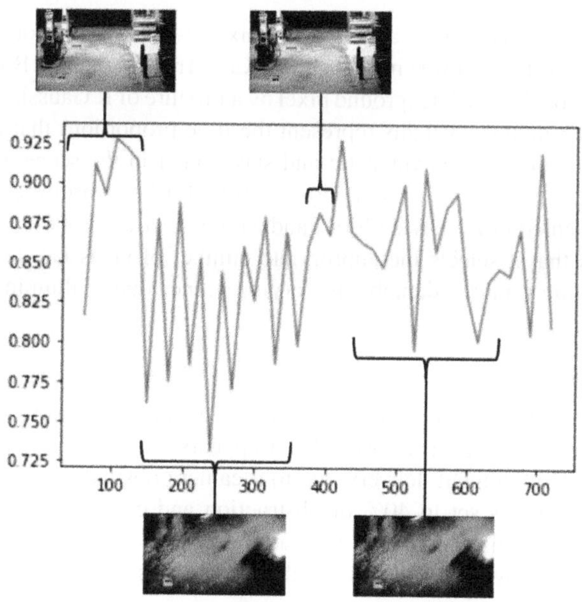

Figure 3.11. *Traditional QoE-VIIDEO score tracking in explosion. For a color version of this figure, see www.iste.co.uk/djeachandrane/smart.zip*

3.3.2.2. *Augmented QoE for content perception assessment*

It is important to consider augmented QoE because even though we achieve good QoE through Video Quality Assessment (VQA), the content may be compromised by voluntary or involuntary obstruction. Camera obstruction is a major issue that refers to any object or activity that blocks or interferes with the view of the camera, making it difficult or impossible to capture clear and accurate footage. This obstruction can be intentional or unintentional, such as placing an object in front of the camera, filming through a dirty or foggy lens, or filming in low-light conditions. In the case of a scene, a leaf that is deposited over the camera could be considered involuntary or was intentionally tampered with by hiding the scene with a truck or another obstructing object. This situation occurs in physical space.

3.3.2.2.1. Proposed method

The proposed method is to detect camera obstruction or tampering. The solution is based on an existing method that involves subtracting the background from the foreground, which allows moving objects to be detected in a space distinct from static objects. In addition, if the foreground occupies a certain percentage of the camera, it is supposed to obstruct or tamper with the camera. Background subtraction is a way of eliminating the background from the image. To achieve this, we extract the moving foreground from the static background.

The proposed method uses a Gaussian mixture-based background/foreground segmentation algorithm that was introduced in KaewTrakulPong and Bowden (2002). This algorithm models each background pixel by a mixture of K Gaussian distributions (K = 3 to 5). The mixture weights represent the time proportions that colors stay in the scene. The colors that remain static and stay longer in the scene are considered probable background colors. The approach was further improved by extended research as presented in Zivkovic (2004) and Zivkovic et al. (2006). One significant improvement is that it selects the appropriate number of Gaussian distributions for each pixel, making it more adaptable to varying scenes due to illumination changes and so on.

3.3.2.2.2. Experiments

In Figure 3.12, the system outputs, in the left window, a normal situation even though the hand is shaken or a novel object appears, and in the right window, the system detects a bad content delivery due to a camera obstruction. This obstruction detection threshold was set to 40% of obstruction and clearly shows its efficiency with this algorithm. This experiment was carried out in a real-world space but is not representative of real-world scenarios and therefore needs to be explored and studied further to make improvements.

3.3.2.3. *Augmented QoE for video integrity assessment*

This QoE is an important aspect to consider when it comes to integrity issues. Even though the content and video quality appear to be acceptable, it is possible that the video flow could be incorrect. For example, the footage may be from the wrong camera, or the original flow might be stuck in a loop. These situations can negatively impact the viewer's experience.

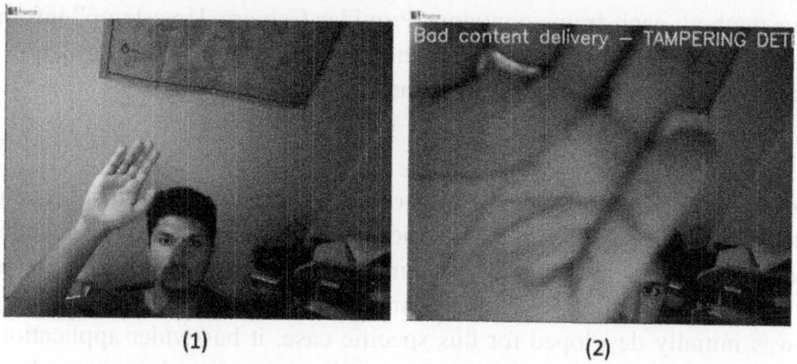

Figure 3.12. *System output – augmented QoE for content perception (source: webcam from computer). For a color version of this figure, see www.iste.co.uk/djeachandrane/smart.zip*

3.3.2.3.1. Proposed method

The proposed method employs a technique called steganography, which is a form of invisible watermarking. Steganography is a technique used to hide information within another message or object in a way that is not noticeable to humans. This provides an advantage over cryptography alone because the hidden message does not draw attention to itself and avoids being scrutinized.

This proposed method is inspired by Juneja and Sandhu (2009) and is based on the Least Significant Bit (LSB) approach and works on lossless compression data formats such as BMP or PNG for digital images, or H.264 for digital videos. However, these formats tend to take up a lot of storage space and can be quite heavy. The main goal of this technique is to hide spatio-temporal watermarking, which contains more precise temporal data information in a textual way within the frames that are being transmitted over the network. This hidden information helps us to identify the authenticity of the data in both spatial and temporal domains. In the spatial domain, we can check if it belongs to the provider system (Airbus DS SLC) and if the flow belongs to the correct camera. In the temporal domain, we can check if the data is temporally coherent, meaning that the data encoded at t+1 comes after the frame encoded at frame t.

For instance, in the case of a tiny 2x2 image, each entry in the array represents a color pixel. The amount of one of the pixel's three primary colors is encoded in each bit triplet (0 or 1), where the rightmost bit of each triplet is known as the low-order bit or LSB. To hide the message 111 111 101 111, we modify the image by setting the LSB of the ith triplet to the value of the ith bit of the message.

In our method, each frame contains “Provider-Camera-Timestamp” information, which is hidden at the camera level using the LBS algorithm. This information is then decoded at the deported intelligence to ensure content integrity.

3.3.2.3.2. Experiments

Figure 3.13 demonstrates the system’s output. The top window displays a normal situation where data integrity is maintained. However, in the bottom window, the system detects a problem with video integrity due to a video loop. This issue arises when a frame at t+1 appears before a frame at t, indicating an error. Although this feature was initially developed for this specific case, it has wider applications. The company tag can be used for authentication, or the camera number tag can be used to ensure the correct flow routing.

(1)

(2)

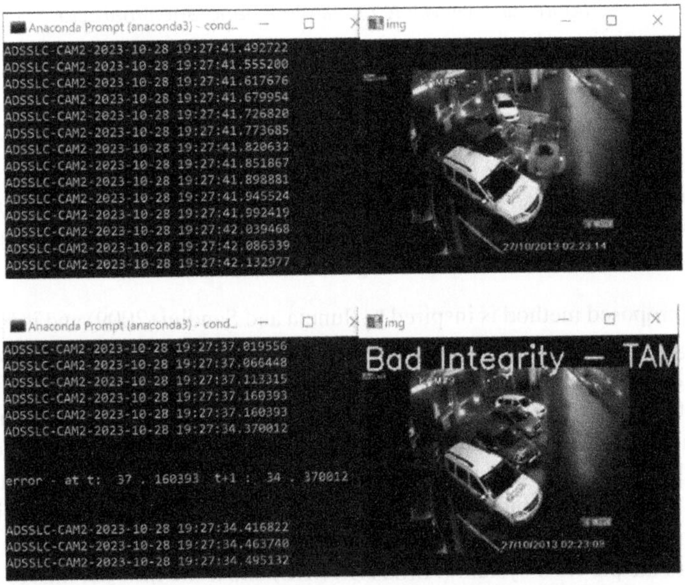

Figure 3.13. *System output – augmented QoE for video integrity. For a color version of this figure, see www.iste.co.uk/djeachandrane/smart.zip*

3.3.2.4. *Task-specific QoE for human preference*

3.3.2.4.1. Proposed method

The reward function R is equal to the task-specific QoE which models human preference as in the following equation:

$$R(s_t, a_t) =_{task-specific} QoE(s_t, a_t) \quad [3.1]$$

where s is the state, a is the action, and t is the discrete time.

The reward function is calculated from the state of the environment as a function of time, thus the task-specific QoE too. It could also be computed from an ensemble τ of states and actions.

The task-specific QoE is equal to the combination of the traditional QoE, augmented QoE and the QoP as follows:

$$_{task-specific}QoE(s_t, a_t) = QoP(s_t, a_t) +_{trad} QoE(s_t) +_{aug} QoE(s_t) \quad [3.2]$$

The QoP is computed with the state and action provided, and the traditional and augmented QoE are computed only with the state.

3.3.2.4.2. Experiments

The experiments will focus solely on the traditional QoE for video quality assessment, in order to draw conclusions about its impact on the QoP. Introducing other augmented QoEs would complicate the experiments due to non-existent obstruction and integrity problems in the dataset.

Two types of noise were added to the video in order to replicate network congestion or noise caused by natural disasters like an explosion. The first noise will be explained in the next section.

In Figure 3.14, blurry noise was added to a sample of number 5 in the MNIST dataset. The degradation of the IQ is shown in order of intensification. It makes it possible to see the impact of noise in signal transmission and how it concretizes in the final rendering. Noise at 1 is the original image without any noise, noise at 5, 10 and 15 are the images altered by noise. At noise 5, the number is readable but at noises 10 and 15, the image is unreadable.

In Table 3.2, NIQE and SSIM for the traditional QoE were assessed and noise was introduced gradually and similarly the loss and the accuracy for QoP is provided. Gradually the data deteriorates, gradually the NIQE and SSIM value decrease. As we can observe, NIQE is not all that accurate compared to SSIM when blurry noise equals 10 and 15. These results demonstrate the limits of the NR approach as NIQE while not having references to compare as in the SSIM one. Nevertheless, it has a certain level of reliability, it generally indicates when the data is little impacted and when it is less so, i.e. readable according to human judgment. The QoP is little impacted, when it was little or not at all affected by noise, the loss was approximately 0 and the accuracy was greater than 0.9 which is not so bad but by adding more noise, the QoP changes drastically, decreasing accuracy and increasing loss. This demonstrates that the exogenous parameters defined in our study clearly have an impact on the endogenous parameters defined in our study. The quality of the data affected the quality of knowledge, and it was assessed through traditional QoE and QoP, which is reliable for future work with fuzzy noise types.

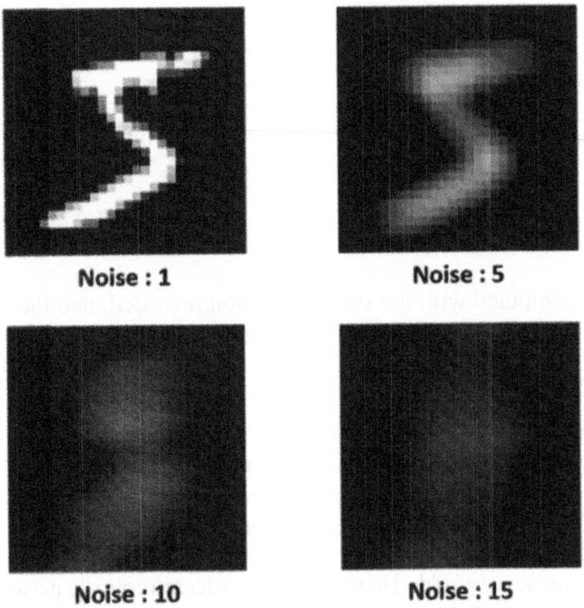

Figure 3.14. *Image degradation by blurry noise on MNIST dataset's number 5*

Blurry noise	1	5	10	15
NIQE (NR-IQA)	133	96	24	26
SSIM (FR-IQA)	1	0.52	0.11	0.03
Loss	0.025	0.466	38.1	123.3
Accuracy	0.99	0.92	0.37	0.06

Table 3.2. *Image degradation impact assessed by QoP and traditional QoE in MNIST dataset with blurry noise*

3.4. Conclusion

The challenge we tackle in this chapter is creating a reliable corrective signal that incorporates enriched contextual information. This signal should encompass the surrounding environment and its contextual awareness to enable effective correction.

In AI, the QoP is used to assess the accuracy of the predictive system for the final task. However, obtaining the ground truth required to compute the QoP is often unfeasible in a non-stationary environment that requires updated knowledge. Obtaining this ground truth is a complicated and unresolved problem in real-world scenarios.

In order to tackle this specific challenge, we initially examined two established methods: direct requests from users and automated labeling. However, we discovered that both of these methods had their own set of issues. The first method was time-consuming, while the second method was not particularly reliable. To address this problem, we explored a hybrid approach, which involved analyzing user data to determine their satisfaction levels. We developed two different methods for this approach. The first method involved filtering out unrelated tasks and using vision-based techniques to assess user satisfaction. However, we found that this method was not suitable for individuals who have a neutral expression. The second method involved combining two different analyses – one based on audio data and the other based on textual data – to minimize uncertainty. While this method yielded promising results, it was difficult to experiment with. We believe that these two approaches could be further extended and applied in the field of video surveillance.

The measurement of QoP alone is not sufficient as it is influenced by many external factors. To ensure system reliability and trustworthiness, the corrective signal should be able to assess the system end-to-end, even in uncontrolled environments. The traditional QoE approach proposed by the computer network community uses a metric called VQA to evaluate video quality distributed in a network with no reference, and it works well. However, this method does not address all of the problems that users may encounter in real-world scenarios. For this reason, we propose two additional augmented QoE methods based on the final task. The first one aims to detect camera obstructions and provide satisfactory application, but it needs to be tested in the final environment. The second method has proven to be efficient in addressing integrity problems. Further research is required for all the methods listed to achieve individual effectiveness and reliability.

In conclusion, combining these approaches in end-to-end, as proposed with "task-specific QoE", will provide overall control over the system and enhance its trustworthiness from the user's perspective.

4

A Situational Awareness-centric Predictive System for Anomaly Detection

"Computers will overtake humans with AI within the next 100 years. When that happens, we need to make sure the computers have goals aligned with ours."
Stephan Hawking

Chapter 4 describes a novel situational awareness-centric predictive system. The purpose is to augment this system's capabilities to meet mission-critical requirements and to reach satisfying results for anomaly detection to operate in real-world situations.

4.1. Context

Machine learning (ML) methods are known as an AI driven by data. Access to the data is the core problem with this type of system. Questions about data frugality arise from this statement, which inquiries about data quality, reliability and availability. Research from this statement on predictive models leads to different topics, such as data quality-aware models, frugal models, and reliable models, where the goal is to build more data-efficient models. In Chapter 3, the data frugality problem was studied for adding context; here, the data frugality problem will be studied to develop a data-efficient model.

For a color version of all the figures in this chapter, see www.iste.co.uk/djeachandrane/smart.zip.

Different challenges will be met, in terms of data and knowledge, which are, as mentioned, data frugality or context-evolving environment problems, for instance, in terms of mission-critical, human factor problems in decision-making, situation awareness problems for reliability, interpretability problems for model truth and real-time problems for real-world applications.

Moreover, our study on a predictive system should match with real-world scenarios not only in terms of application but also in terms of implementation, on existing video surveillance system implementations. Among existing solution deployment today in video surveillance, problems when deploying an adaptive system could appear, such as a labeling process when it comes to a multi-camera video surveillance system or when a long sequence of video is gathered and should be annotated.

In scenario 1: multi-camera video surveillance system

When many cameras are used in a monitoring room and one user is supervising, the camera detecting the abnormal situation should be highlighted, and for this camera, if the abnormal situation is true or not, the feedback signal should be provided to the system respectively to the correct camera. In and for an adaptive system, it could be a complex task to find the correct camera and provide labels to data collected only from this camera.

In scenario 2: video frame and sequence analysis

When video data is captured and analyzed in a control room, the feedback signal could be time-consuming and labor-intensive if being built at each frame.

Both of these scenarios are impracticable and would render the system inoperable.

4.2. Baseline

The dataset description is provided as follows.

– Name: UCF-Crime.

– Duration: 128 hours of video.

– Anomalies: abuse, arrest, arson, assault, burglary, explosion, fighting, traffic accident, theft (victim present), shooting with a firearm, shoplifting, theft (without the presence of the victim), vandalism.

The baseline framework is presented in Figure 4.1.

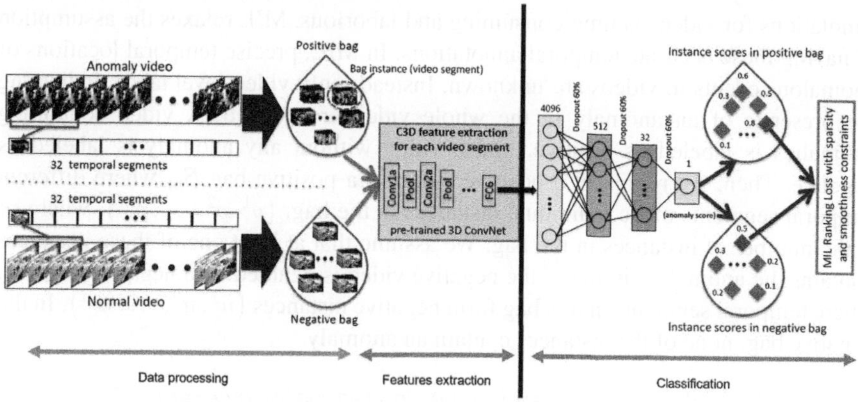

Figure 4.1. *Baseline framework*

The data processing and feature extraction consist of the following.

– Cut video 16 frames to 16 frames.

– Feature extraction with C3D for each segment.

– Build new segment by computing mean of segments to create 32 new segments per video.

– Load the batch, 30 random abnormal videos and 30 random normal videos.

Feature extraction was made by a neural network (NN) called CNN-3D.

The classifier is multi-layer perceptron which uses a custom loss function also known as similarity loss. The function loss is a derivative of triplet loss where only max-min instance value of a video sequence is chosen for the optimization. And a term for sparsity and a small temporal constraint is fixed to define relationships between instances in the same sequence.

4.3. Problem statement

In our work, four problems should be studied and solved as a whole, individually, or partially.

Problem 1 – temporal anomaly localization, i.e. "credit assignment"

To learn a robust classifier, accurate annotations of positive and negative examples are required. In the context of supervised anomaly detection, a classifier needs temporal annotations of each segment in videos. However, obtaining temporal annotations for videos is time-consuming and laborious. MIL relaxes the assumption of having these accurate temporal annotations. In MIL, precise temporal locations of anomalous events in videos are unknown. Instead, only video-level labels indicating the presence of an anomaly in the whole video are needed. A video containing anomalies is labeled as positive, and a video without any anomaly is labeled as negative. Then, we represent a positive video as a positive bag $\mathcal{B}_p$, where different temporal segments make individual instances in the bag, $(p^1, p^2, \ldots, p^m)$, where m is the number of instances in the bag. We assume that at least one of these instances contains the anomaly. Similarly, the negative video is denoted by a negative bag, $\mathcal{B}_n$, where temporal segments in this bag form negative instances $(n^1, n^2, \ldots, n^m)$. In the negative bag, none of the instances contain an anomaly.

Problem 2 – trial-and-error learning, i.e. "no teachable by example"

In the context of RL anomaly detection, a classifier does not have annotations. RL relaxes the assumptions of having prior knowledge on the bag's annotations. In RL, only T trajectory containing a couple of S states $(s_t, s_{t+1}, \ldots, s_{T-1})$ as in Figure 4.2 and A actions $(a_t, a_{t+1}, \ldots, a_{T-1})$ is provided. And then, the environment will return the reward R, and by combining A actions and R reward information, ground truth could be retrieved in binary classification as for anomaly detection.

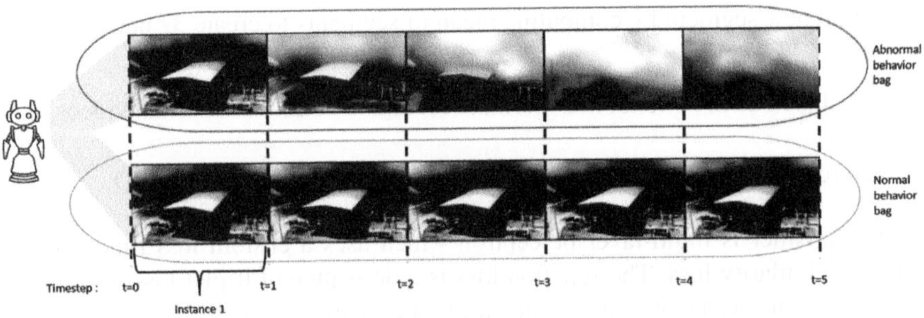

Figure 4.2. *Reinforcement learning as a multi-instance learning framework: trajectory as a bag and a time step as an instance*

Problem 3 – concept drift

As stated in Verdecia-Cabrera et al. (2019), in online learning, a classification task is generally defined for a sequence (possibly infinite) of instances S = e1, e2, ..., ei, ... arriving over time. Every training instance $e_i = (\vec{x_i}, y_i)$ is formed by a vector $\vec{x}_i$ and a discrete value y_i. Each vector $\vec{x_i} \in \vec{X}$ has the same dimensions. Each dimension is named attribute, and each component $x_{ij} \in \vec{x}_i$ is an attribute value (numeric or symbolic). The discrete value y_i named label and taken from a finite set Y of possible class values. It is commonly assumed that the data stream S is generated by a probability density function $P(\vec{X}, Y)$. The classification learning task is to obtain a model from S that approximates P as $\hat{P}$ so that $\hat{P}$ maximizes the predictive accuracy. Concept refers to the probability distribution function of the problem at a given timestamp. Therefore, a change in P after a timestamp entails a concept change or concept drift. Gama et al. (2014) distinguish two main types of concept drift.

– Real concept drift refers to changes in the distribution of posterior probability of the classes $P(Y|X)$. These changes can occur without a change in the probability distribution of the instance space P (X). It is strongly related to the prior knowledge known as labels.

– Virtual concept drift happens when the probability distribution of the instance space changes P (X). It purely refers to data.

Problem 4 – model interpretability

Models are described as equations or formulas. In general, there is a dependent variable (y variable) expressed as a function of independent variable(s) (x variable) with various constant and/or parameters:

$$\mathbf{y} = \mathbf{f}(\mathbf{x}, \mathbf{p}, \mathbf{c}) \qquad [4.1]$$

where y is the dependent variable (observed, calculated), x is the independent variable, p is the adjustable parameters, and c is the constants which is fixed (duration, etc.).

In deep learning, the variable P plays a crucial role in the functioning of the model. However, this variable can be difficult to understand and when that happens, it is referred to as a "black box". By interpreting the parameters that represent the inner model, we can gain insight into how the model works and improve our ability to make accurate predictions. This increased understanding can lead to more trust in the model's capabilities and better outcomes in various applications. Unfortunately, there is no consensus over the definition of interpretability, as explained in Laugel (2020). Therefore, the aim will be to interpret the reasoning behind our proposed mechanism to resolve Problem 1.

4.4. Proposals

First, in brick reasoning, pre-processing and feature extractor will be studied, and then classifiers will be studied in depth. Our focus will be on improving classifiers, which is a challenging task due to various issues we face, as well as the scientific opportunities and tools available to explore potential solutions. Our goal is to address these challenges and contribute to finding answers.

4.4.1. *Feature extraction experimentation and reviewing*

For our study, two methods based on existing work done in anomaly detection on UCF-Crime were selected, and we chose the best approach in terms of anomaly detection level application and catching temporal patterns.

In the paper by Sultani et al. (2018), they first introduce the dataset; researchers work with the 3D CNN extractor to extract spatial and temporal features and classify them with a multi-layer perceptron.

In a second work by Anala et al. (2019), they use a 2D CNN to extract spatial features and a long short-term memory (LSTM) as a sequence learner; this method provides a good accuracy, but the area under the curve (AUC) is not mentioned. The LSTM was trained for action recognition task only on a subset of the full UCF-Crime dataset, i.e. Normal, Road Accidents, Fighting and Explosion. For our work, we reproduce this feature extraction architecture but with the full dataset and following the data pipeline presented above.

Ref.	Detection level		Result (AUC %)	
	Mentioned	Obtained	Mentioned	Obtained
Sultani et al. (2018)	Segment-level	Segment-level	75	77.5
Anala et al. (2019)	Video-level	Frame-level	-	66

Table 4.1. *Comparison study of feature extraction between 3D-CNN and LSTM and 2D-CNN*

The reimplemented methods shown in Table 4.1 allow us to validate the approach based on results to move towards our future implementation. Methods presented by Sultani et al. (2018) and Anala et al. (2019) that show the temporal patterns in video anomaly detection are processed in two different ways. In Sultani et al. (2018), the author uses 3D features as input and classification is done segment by segment, while in Anala et al. (2019), 2D features are used as input and classification is done on frame sequences, but this second model was not correctly adjusted. Hence, by experimenting, we selected the first feature extraction implementation.

To replicate the experiments, the exact configuration details from the baseline paper were followed, but the reimplementation was done on different libraries due to our desire to harmonize all codes with the same library, TensorFlow (Abadi et al. 2015). By experimenting, the result gap noticed with the baseline approach is only due to the library's differences. In future comparisons with the scientific literature, we will reference the 75% AUC mentioned in their article to ensure consistency with other researchers. However, for further experiments aimed at improving results, we will use the approximately 78% AUC obtained from our initial experiment as a benchmark in our experimental plan.

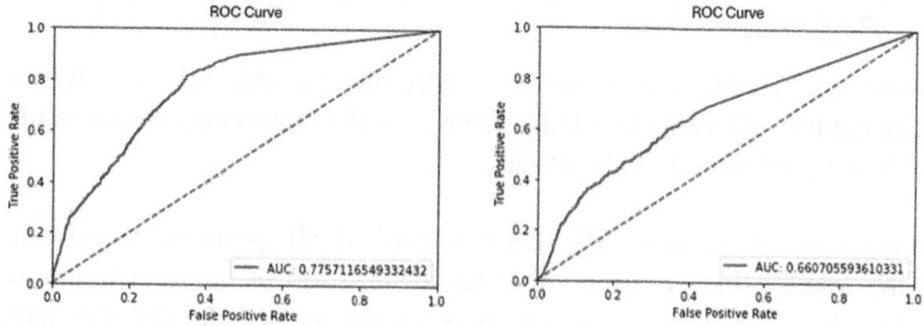

Figure 4.3. *Comparison between C3D and LSTM/CNN feature extractor*

In addition, according to Table 1.11 in the literature review of methods applied to the UCF-crime dataset for anomaly detection, the C3D feature extractor was chosen because it was used five times in our reviewed methods, which allows for a deeper analysis to understand the MIL classifier independent of the feature extractor. And further, we can consider the impact of other chosen Extracted Features (EFs) in recent studies; it has its impact, especially I3D (Carreira and Zisserman 2017) versus C3D (Tran et al. 2014) and TSN (Wang et al. 2016) versus I3D (Carreira and Zisserman 2017). Considering the impact of Feature Extractor (FE) could be interesting in the production phase independently of the classifier. The use of a specific FE shows the limits of the classifier which obtained good results only thanks to a better FE and not to a better MIL classifier.

4.4.2. *Capability-oriented classifier study*

To answer mission-critical properties, two approaches will be presented, which were not explored in our literature review for abnormal situation detection. The first one is the need for an interpretable model, and the second one is the need for an adaptive model.

4.4.2.1. *Interpretable model – self-organizing map and hidden Markov model*

By definition, models are interpretable when humans can readily understand the reasoning behind predictions and decisions made by the model. The more interpretable the models are, the easier it is for someone to comprehend and trust the model. To get the difference between interpretability and explainability, interpretability has to do with how accurately an ML model can associate a cause to an effect. Explainability has to do with the ability of the parameters to justify the results. This area of research was raised when knowing that ML algorithms have historically been "black boxes" that provided no way to understand their inner processes and made it difficult to explain resulting insights to regulatory agencies and stakeholders, which is more relevant in the context of mission-critical operations. As explained in Problem 4 of section 4.3, we wish here to understand the mechanisms that will be developed to solve Problem 1.

4.4.2.1.1. Proposed method

To answer this problem, first was to explore interpretable ML, which means that humans can capture relevant knowledge from a model concerning relationships either contained in data or learned by the model.

To proceed to interpretable ML, using statistical ML permits the explanation of data and the relationship between variables in the form of mathematical equations. Incidentally, these models use a lot of assumptions to identify the underlying distribution and relationships. These models use mathematical equations to make predictions and have a clear understanding of how to interpret the parameters, which can aid in determining how the data relate to one another.

Amongst statistical ML models, we were looking for modeling which is able to capture spatial and temporal dependencies. And then looking for a possible implementation to do multi-instance learning. For that, we will investigate the hidden Markov model (HMM) (Baum and Petrie 1966) and the self-organizing map (SOM) (Kohonen 2004) in chain.

HMM

For that, the HMM was the method that was the most suitable for our application. A specific architecture was developed to perfectly suit our use case. HMM is a statistical Markov model in which the system being modeled is assumed to be a Markov process with unobserved “hidden states”. In probability theory, a Markov model is a stochastic model used to model randomly changing systems. It is assumed that future states depend only on the current state, not on the events that occurred before it (i.e. it assumes the Markov property). HMM answers three problems as follows.

– Evaluation problem: What is the probability that a particular sequence of symbols is produced by a particular model?

– Decoding problem: Given a sequence of symbols (your observations) and a model, what is the most likely sequence of hidden states that produced the sequence?

– Training problem: Given a model structure and a set of sequences, find the model that best fits the data.

This third problem makes it possible to consider this statistic model as ML one. Then, statistical interpretation could be done by evaluation or decoding problems.

The HMM is characterized by three features: the prior probability of the start state, the emission probability and the transition probability. The HMM is composed of two types of sequence: the Markov States Sequence, also known as the Hidden State or latent space in machine learning, and the emission sequence, referred to as the observation state.

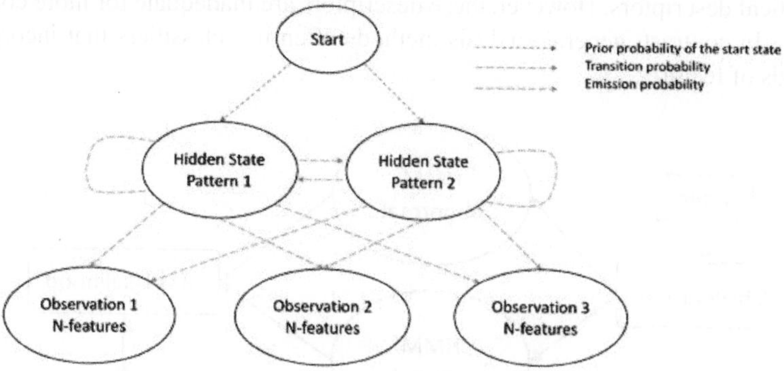

Figure 4.4. *HMM statistical model*

As in Figure 4.4, in our case, the number of hidden states should be empirically studied because it will be interpreted as an inner pattern to be modeled.

For our approach, we will use HMM to resolve two problems: learning and evaluation for the inference part. For that, we will use two HMMs, one to learn from abnormal and normal data and the other one to learn from only normal data. Then, we will evaluate by computing the likelihood, the probability of data produced by the model.

For each problem, we have implemented the following.

– Evaluate: compute the probability of the observation sequence (Forward Algorithm).

– Learning: unsupervised training of HMM parameters (Baum–Welch Algorithm).

Knowing that we have two models, the model which obtains the highest likelihood will be selected. For the MIL application, the HMM will be inferred from one observation for the classification; in this case, when using the forward algorithm, the prior probability of the start state has a major impact on the computing of the likelihood while no recursion is proceeded in the inference phase, so transition probability will not impact the inference stage. The HMM classification scheme used is as in Figure 4.5.

Although this method was adapted for our use case, one of the major drawbacks after many years of application is that HMMs are limited in their ability to capture only linear relationships between observations and states. Furthermore, they are sensitive to the choice and quality of the features and parameters which may affect their performance. Our literature review highlights the application of Hidden Markov Models (HMM) in crowd scene analysis, which typically relies on a limited number of statistical descriptors. However, these descriptors are inadequate for more complex analyses. In contrast, general analysis methods often use classifiers that incorporate thousands of features.

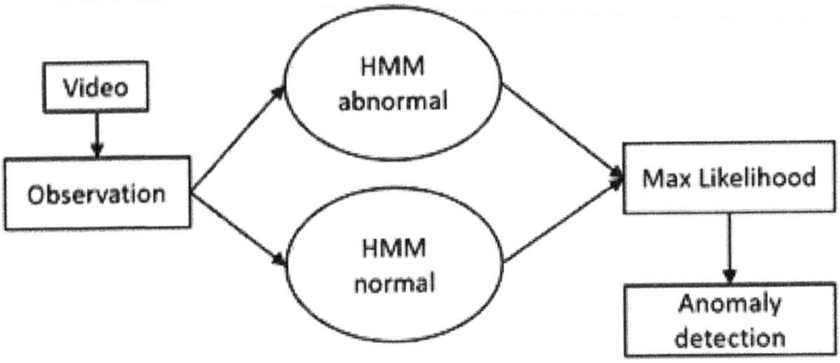

Figure 4.5. *HMMs anomaly detection framework*

To maintain an interpretable model while addressing this issue, we implemented the Self-Organizing Map (SOM) algorithm for dimensionality reduction. This approach serves as a filter, providing interpretable new descriptors for our analysis.

SOM

SOM makes it possible to map states from an uninterpretable continuous space to a lower-dimensional space with a predefined topologically interpretable structure, such as an easily visualizable two-dimensional grid. These clusters could then be visualized as a two-dimensional "map" such that observations in proximal clusters have more similar values than observations in distal clusters. This can make high-dimensional data easier to visualize and analyze. An SOM is a type of artificial NN but is trained using competitive learning rather than the error-based learning used by other artificial NNs.

SOM is able to capture nonlinear relationships and therefore SOM forms a semantic map where similar samples are mapped close together and dissimilar ones apart. This may be visualized by a U-Matrix (Euclidean distance between weight vectors of neighboring cells) of the SOM.

The training utilizes competitive learning. When a training example is fed to the network, its Euclidean distance to all weight vectors is computed. The neuron whose weight vector is most similar to the input is called the best matching unit (BMU). The weights of the BMU and neurons close to it in the SOM grid are adjusted towards the input vector. The magnitude of the change decreases with time and with the grid distance from the BMU.

The specificity of this method is that its NN architecture is its feature map architecture. And the codebook vector is the dimensionally reduced feature containing the weight matrix between the input vector and feature map.

For the training of our SOM model, we train it on abnormal data which contain a mixture of normal and abnormal data. And then, it permits ingesting 4,097 features of C3D and provide 100 new features.

HMM and SOM

For our proposal shown in Figure 4.6, we developed an end-to-end interpretable classifier which reduced dimensions with the SOM and use the HMM to classify the reduced data dimension to detect abnormal situations.

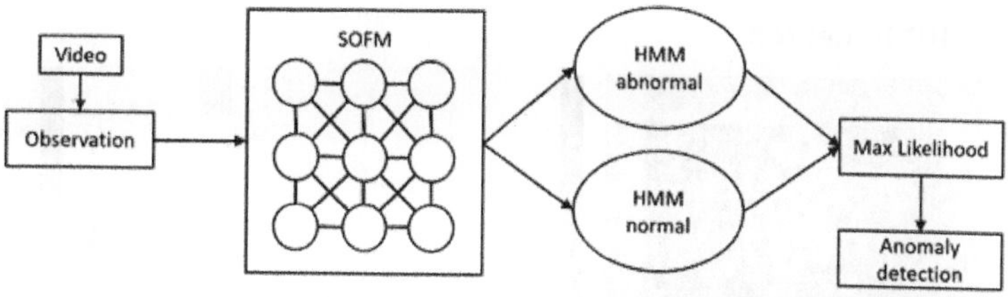

Figure 4.6. *HMM and SOM-based anomaly detection framework*

4.4.2.1.2. Experiments

SOM study

For our experiment, training was done in an unsupervised way thanks to the competitive learning algorithm over the abnormal dataset of UCF-Crime, as in Figure 4.7. First, an NN with 100*100 neurons was tested and then with 10*10 neurons. For our experiment, the U-matrix was shown which is the distance map of what was learned, and the activation map was also shown which is the BMU activation map over 100 samples.

The U-matrix makes it possible to study patterns and clusters identified in data, and the activation map could show the cluster by activating the BMU. The U-matrix of high-dimensional SOM shows clusters that are light colors surrounded by dark ones. But unfortunately, this dimension stays high according to how HMM could be used. To address this, we use a lower-dimensional SOM. While the U-matrix is challenging to interpret due to some extremely dark neurons, the activation map is clearer, showing the most activated neurons. At this stage, the lowest representation needs a deeper study of the activation map. The SOM activation map was computed on a full abnormal set and a normal set separately in order to find dissimilar patterns as in Figure 4.8.

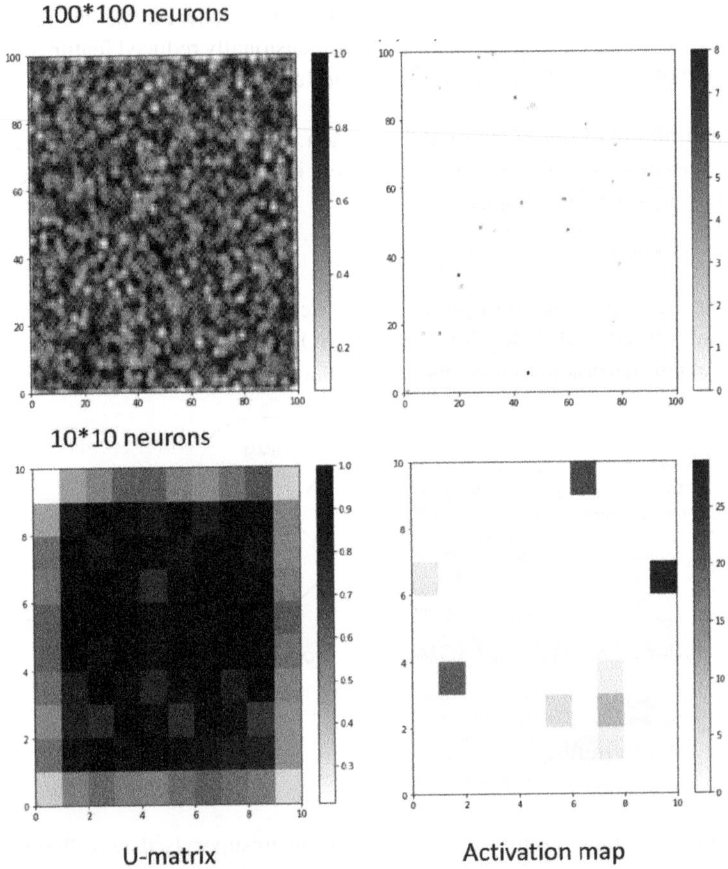

Figure 4.7. *U-matrix and activation map of SOM with 100 neurons and 10,000 neurons*

At this stage, we proceed to use a simple probability reasoning that considers in the anomaly detection problem normal and abnormal situations as a mutually exclusive event. Mutually exclusive events are two or more events that cannot occur at the same time. Mutually exclusive events are sometimes called disjoint events. If two events are mutually exclusive, then

$$P(A \text{ or } B) = P(A) + P(B) \quad [4.2]$$

This means that the probability of event A or event B occurring is equal to the probability of event A occurring, plus the probability of event B occurring.

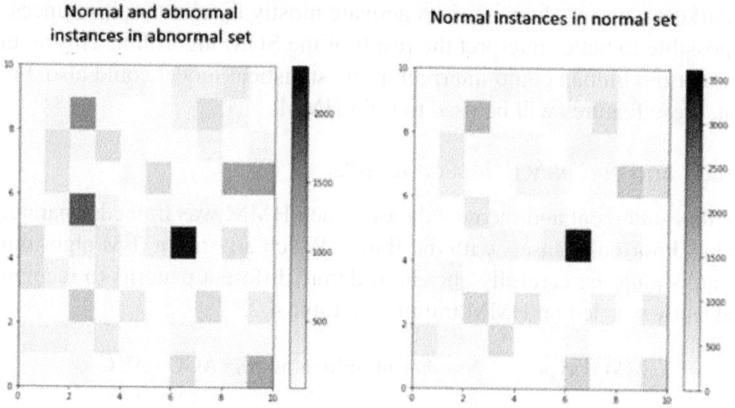

Figure 4.8. *Activation map comparison between abnormal and normal sets*

Fortunately, the SOM would be understood as a discrete approximation of the distribution of training samples, with neural weights acting as pointers to the input space. More neurons point to regions with high training sample concentration and fewer where the samples are scarce. Then, by this assumption, the abnormal set containing a normal and abnormal instances activation map will be minus by the normal set containing only normal instances in order to obtain only an abnormal instances activation map, as in Figure 4.9.

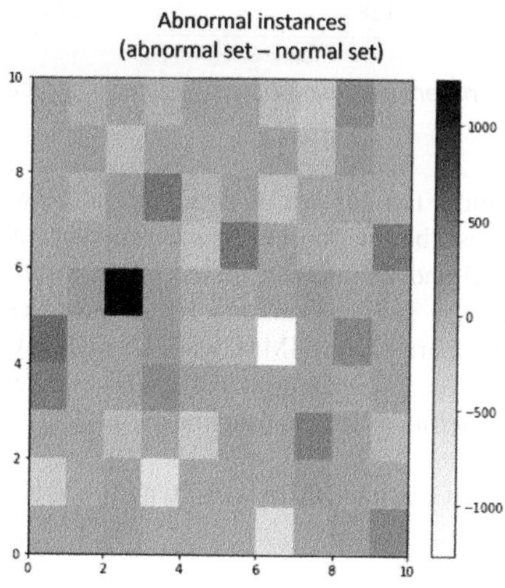

Figure 4.9. *Activation map of abnormal instances*

The darkest neurons show neurons activate mostly by abnormal instances, which make it possible to better interpret the result of the SOM algorithm. This assumption supposes that if a human could interpret it, the statistical model could also. From this statement, these features will be used to train HMMs.

HMM study and anomaly detection results

From this statement and the new dataset, each HMM was trained separately from abnormal and normal datasets with the Baum–Welch algorithm (EM algorithm). The hidden state should be carefully chosen to define different patterns to recognize. An empirical study was led on HMM training in Table 4.2.

HMM type	Number of hidden nodes	ACC	AUC
	2	0.855	0.46
	4	0.786	0.57
	8	0.80	0.61
GaussianHMM	16	0.82	0.65
	32	0.82	0.62
	64	0.83	0.60
	128	0.83	0.56
	2	0.85	0.46
	4	0.78	0.57
	8	0.80	0.61
GMMHMM	16	0.81	0.64
	32	0.83	0.60
	64	0.85	0.60
	128	0.78	0.56

Table 4.2. *HMM experiment over distribution type and number of hidden nodes*

For the experimentation, the number of nodes varied between 2 by multiplying by two until 128. And distribution chosen for each hidden state to generate data is a simple Gaussian model and a Gaussian mixed model containing, by default, 10 components. Surprisingly, a simple Gaussian HMM provides a better result than the mixed one when compared with HMM with 16 nodes. Once trained, the full pipeline was inferred from the testing dataset and was assessed at frame-level anomaly detection with the AUC metric, as for the baseline method.

The better model is the Gaussian HMM with 16 nodes, as shown in the ROC curve in Figure 4.10.

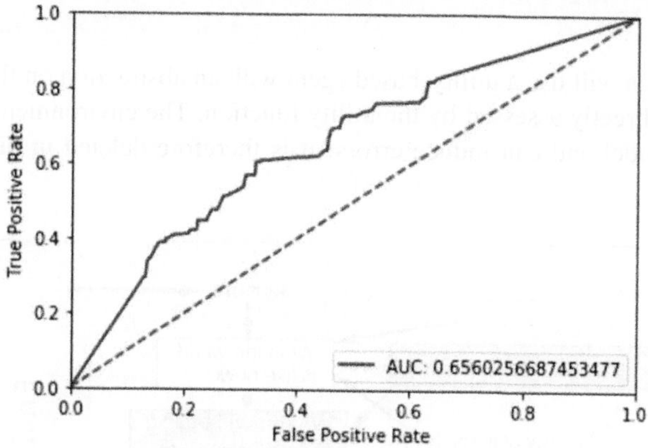

Figure 4.10. *Best HMM model AUC for frame-level anomaly detection*

As we can see, it is giving satisfying results and good improvement in terms of interpretability according to existing methods reviewed in the literature for video surveillance in general and for complex ones, too. Unfortunately, potentially due to strong assumptions in statistical modeling, results in terms of accuracy are not satisfying enough at this stage.

This approach has the advantage of using SOM for spatial pattern recognition without supervision and HMM for temporal pattern recognition in a supervised way. Both are standard ML models that are less greedy than the deep learning approach, which means this approach can be considered data efficient.

4.4.2.2. *Adaptive model – Actor-Critic reinforcement learning*

The adaptive model is important in mission-critical operations when it refers to knowledge plan management, which is crucial in an ever-changing environment and a context-evolving environment. Time-frozen knowledge could be obsolete in real-world scenarios and could lead to critical situations. For this, a time-evolving system makes it possible to update the knowledge gradually over time. At this stage, a simple model differs from an agent. An agent is an entity that perceives and interacts with its environment. An anticipating model could make decisions by having future insight. It could be important in mission-critical operations by forecasting future information to take a decision at the current moment. Having insight into the future allows us to make relevant decisions in order to prevent a situation and allows first responders to anticipate critical situations.

4.4.2.2.1. Proposed method

First, the agent will be designed, then the type of ML methods will be chosen, and finally, the model and algorithm will be defined.

Agent

Our approach will use a utility-based agent with an abstraction on the model part, which will be directly assessed by the utility function. The environment model could be tricky to model and can induce errors; it is therefore deleted in our agent as in Figure 4.11.

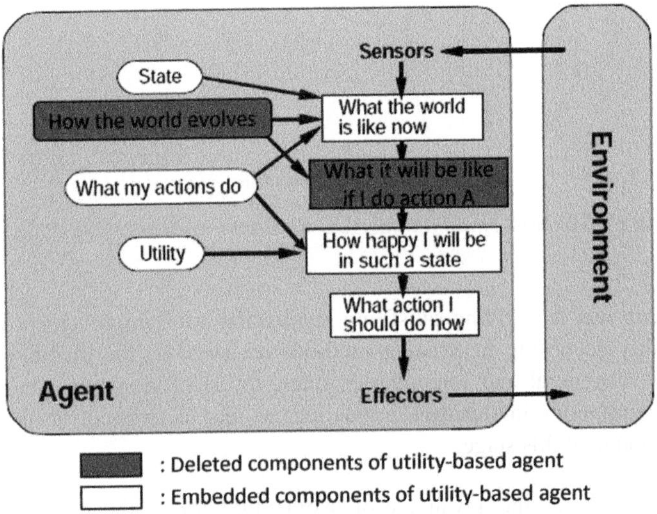

Figure 4.11. *Our intelligent agent architecture inspired by a common utility-based agent*

RL paradigm

In this part, the choice of the paradigms depends on the presence of historical data and/or the complexity of the prior knowledge. Even though we have a historical dataset, in real-world scenarios it will not matter if the context evolves, so the complexity of prior knowledge will be important at this stage. The prior knowledge corresponds to the corrective signal related to the perceived data. This prior knowledge allows us to define the labeling complexity, which differs from label complexity. Label complexity is the fact of giving an input to each data, the labeling complexity is how to build this input for the data. The simplest way of labeling is to proceed to trail-and-error reasoning as humans do. On the contrary, label complexity could be resolved by an unsupervised method, but it will not have supervision and could not be adaptive as expected by human supervision. For that, RL methods are chosen to preserve human supervision and reduce labeling complexity for time-evolving knowledge. Reinforcement learning (RL) is defined by its environmental feedback and its ability to deal with no past experience by gathering knowledge through environmental interaction.

Feedback signal reduces the labeling complexity, and the learning process goes through environmental interaction over time, which permits continuous learning.

To overcome the above-stated challenges, among the main ML paradigms, we will use RL for two main reasons:

– dealing with incomplete data;

– reducing labeling complexity.

RL is one of the three basic ML paradigms, where the training method is based on rewarding desired behaviors or punishing undesired ones. The algorithm aims at enabling an agent to perceive and interpret its environment and to take actions through trial and error. Deep RL is the combination of deep learning and RL in order to map more complex relationships to learn a policy.

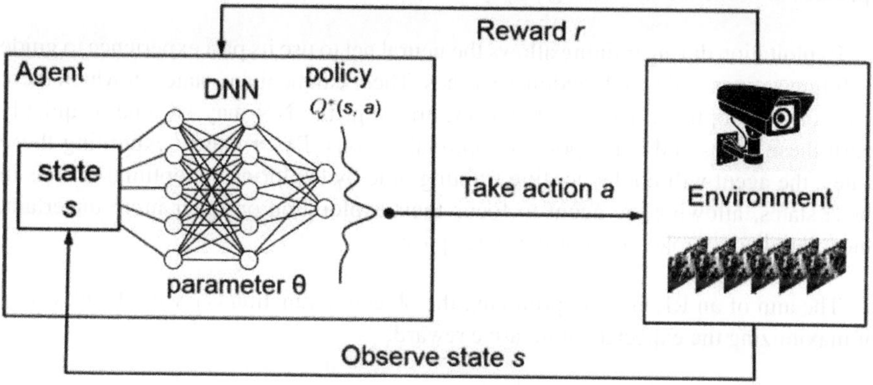

Figure 4.12. *RL for anomaly detection on surveillance videos*

In CCTV footage, our anomaly detection problem is solved by making the anomaly detection algorithm an RL task that considers the reward as feedback from the environment. The RL model contains agent, state, action, reward and policy that operate in an environment. In Figure 4.12, the details are described as follows.

– Environment: CCTV system.

– Agent: a learning entity in the DSS.

– State: a frame/segment of the video.

– Action: an agent's reaction to observed videos resulting in a categorization of normal and abnormal scenes.

– Policy: relation between states and actions derived from the predictive function.

– Reward: feedback to the agent.

In fact, whatever the action taken by an agent from a state, a reward will be given to them by the environment and it gives information about how this action was: positive or negative. RL aims to maximize this reward. The higher it is, the better the agent learns. It has its own learning process via exploration and exploitation.

An algorithm that chooses to always explore during training is unlikely to find an optimal policy because it will employ a more random search as opposed to a directed search. During training, the NN aims to determine the relation between states or state-action pairs and the reward signal through past experience. If the agent is always exploring during training, it will never use the experience gained from past episodes to influence its training policy and, therefore, will search more randomly.

Exploitation during training allows the neural net to use its past experience to guide its future actions and avoid random searches. There can be many states in which there is an obvious optimal action. After some training, the NN may be able to quickly learn these states and corresponding optimal actions. By primarily exploiting those states, the agent will not be wasting training time by exploring suboptimal actions in those states, allowing the agent to focus their exploration on other more uncertain, unexplored or complex parts of the state space.

The aim of an RL task is optimizing the objective function O_f with the objective of maximizing the expected cumulative reward:

$$O_f = \text{Max}\, E\left[\sum_{t=0}^{\infty} \gamma^t \times r_t\right] \qquad [4.3]$$

where $\gamma \in [0, 1]$ is the discount factor, r is the reward, and t is the discrete time.

In deep RL, the policy is adjusted approximately using NNs. NNs are used in addition to standard RL when the latter alone cannot deal effectively with high and complex observation space, which could be represented in CCTV by two- or three-dimensional data.

Actor-Critic algorithm

In addition, to overcome the above-stated challenges, we will use Actor-Critic RL for two main reasons:

– online adaptive learning for updating SA knowledge in near real-time;

– value function for inferring in the present from future information.

For further details, the pros and cons of our approach are highlighted in Table 4.3.

Key elements		**Pros**	**Cons**
Online adaptive learning	**Adaptive learning**	Continuously evaluates and corrects errors made by the classifier. Ability to learn the latest concept and move away from the obsolete model. Handles new types of anomalies to cope with open-world classification	Building a signal is labor-intensive work and making it instantly available can be restrictive
	Online learning	Adapts to fast changes in data. Allows fast model updates in a non-compute-intensive way	Sequential learning is time-consuming for model generalization
	Labeling complexity	Evaluative feedback reduces labeling complexity by only informing if the system is returning good insight	Difficult to readapt the reinforcement signal to other frameworks needing labels
Projection	**Memory-less forecasting**	Reduces the spatial complexity when predicting the value-function through the Markov property	The hypothesis introduces some limitations to catch temporal patterns
	Anticipating	The non-myopic agent infers to the present from future insight	Error computing is time-consuming

Table 4.3. *The pros and cons of our approach*

In RL, several algorithms exist. The choice of the algorithm that suits our use case depends on many factors. We pay special attention to environment dynamics, real-time requirements, data management efficiency and uncertainty modeling. The chosen algorithm, Actor-Critic (Bhatnagar et al. 2009), answers theoretically to all these aspects.

– **Environment dynamics**: a model-free method is preferred to a model-based method due to unknown information about the Markov Decision Process (MDP) associated with the environment because of the aleatory properties of the environment. The agent will need to estimate a policy by reinforcing their beliefs about the dynamics of the environment.

– **Real-time requirements**: with Monte Carlo (MC) learning, the state value function cannot be updated until the end of the episode. Therefore, it is difficult to learn in real time. Hence, time difference (TD) learning is considered because it can update policy/value at each step, which makes it suitable for real-time applications.

– **Data management efficiency**: TD learning includes a naturally bootstrapping method, unlike MC learning. In fact, in TD learning, returns are obtained by estimating future discounted rewards, while in MC they use the exact returns. Using longer trajectories without bootstrapping can often lead to high variance, which, in practice, means we need more samples before the estimates converge. High variance signifies that it will not generalize on previously unseen data. Bootstrapping in RL means that we update a value from estimations and not on exact values, which makes it possible to reduce the variance.

– **Uncertainty modeling**: due to the complexity of environments, it is important to model this uncertainty through a stochastic policy. The most appropriate solution for this is to use a policy-based method that takes into account uncertainty related to actions rather than V values.

An important point to highlight is that the policy-based method uses MC learning, which causes problems in operating in real time, which is why an Actor-Critic hybrid approach is chosen to deal with uncertainty and real time.

Actor-Critic method for SA levels 2 and 3

The Actor-Critic approach consists of adopting a stochastic policy and TD learning. To achieve this, we need two entities – one to predict the possible action with a probability distribution and the other which permits predicting the V value function at each step in order to correct the actor by giving information about how good this state is. And together, they can perform in real time, as shown in Algorithm I.

In Algorithm I, two-stage optimizations occur to proceed to a classification by taking future insight into account. The losses to minimize are related to their tasks.

One entity, "Critic", tries to maximize the expected future reward by computing mean square error loss, and the second entity, "Actor", tries to discriminate actions to choose from by performing binary cross-entropy in the direction suggested by the critic. The critic provides the TD error to itself and to the actor. It is used to optimize the agent's policy and value functions. This signal will first help the critic to better reward forecasting and will encourage the actor to follow the direction of the critic. This method is called reward-based updating. A positive TD error increases the probability of selecting an action in the current state, while a negative TD error decreases it. From the Bellman equation, TD error is obtained by calculating the difference between two different ways of estimating the same value, where one of them, the TD target, is expected to be a slightly more reliable estimate because it is based on the current reward. The actor will then be able to take the action, abnormal or normal, depending on reward forecasting.

Algorithm 1: TD Actor Critic

Randomly initialize critic network $V_{\pi}^{U}(s)$ and actor network $\pi^{\theta}(s)$ with weights U and θ
Initialize environment E
for *episode=1,M* **do**
  Receive initial observation state s_0 from E;
  for *t=0,T* **do**
    Sample action $a_t \sim \pi(a \mid s)$ according to current policy
    Execute action a_t and observe reward r and next state s_{t+1} from E
    Set TD target $y_t = r + \gamma \cdot V_{\pi}^{U}(s_{t+1})$
    Update critic by minimizing loss:
      $\delta_t = \left(y_t - V_{\pi}^{U}(s_t)\right)^2$
    Update actor policy by minimizing loss:
      $\text{Loss} = -\log\left(\pi^{\theta}(a \mid s)\right) \cdot \delta_t$
    Update $s_t \leftarrow s_{t+1}$
  end
end

4.4.2.2.2. Experiments

In our previous study (Djeachandrane et al. 2022), we chose the best approach in terms of anomaly detection-level application and catching temporal patterns among three methods (Sultani et al. 2018; Aberkane and Elarbi 2019; Anala et al. 2019) based on existing work done in anomaly detection on UCF-Crime. By reimplementing and experimenting, we selected the C3D feature extractor (Tran et al. 2014). The experiment details of data-preprocessing and feature extraction can be found in Sultani et al. (2018).

Here, in RL, the agent needs an environment to interact. Our simulated environment provides real-time segments with 4,096 features in a sequential way. All videos present in UCF-Crime are decomposed into 32 segments and were used as input for the environment. The environment provides data in a sequential way, and in order to break the relation between the last segment of a video and the first segment of the next video, we randomize the order each time the environment is finished. These segments were fed to the agent when the latter interacted with the environment. In the trial-and-error process, the environment returns the evaluative reward if it was well classified or not. In our experiment, the evaluative feedback was also simulated from the original labels. It allows for performing adaptive control over the classifier thanks to a corrective signal. We allow our agent to work on a real-time data stream by simulating a real-time environment, and this data stream allows our agent to collect and ingest data step by step to update the model. Next, the Actor-Critic agent classifies a segment as normal or abnormal. Finally, we extrapolate segment scores linearly to compare with the frame-level ground truth by considering 0.5 as the threshold. Above this value, it is considered abnormal, and below, it is considered normal.

Two analyses were processed – execution time and anomaly detection. These two assessments allow us to show their application in real-world scenarios.

Configuration details

The actor and critic were modeled using one NN with common layers and two separate output layers. The NN is set as follows: the first common layer uses 512 neurons with ReLU activation, the second common layer uses 32 neurons with ReLU activation, the first output layer, "actor", uses two neurons with softmax activation, and the second output layer, "critic", uses one neuron with linear activation.

The learning rate α and gamma γ were respectively set to 0.001 and 0.99. In the training loop, the maximum number of episodes M was fixed at 500 episodes and the optimizer used is Adagrad. The GPU used for our experiment was the NVIDIA Quadro P4000 with 8 GB vRAM.

Execution time results

Time-wise analysis is important to assess the ability of the algorithm to deal with a real-time environment. Table 4.4 shows execution times taken by the online adaptive learning algorithm. This study was done on one example, Abuse_28 which is a video of 47 seconds with a frame rate of 30.

Phase		**Time (s)**
Online adaptive learning	**Inference**	0.0024 /seg 0.0448 /32seg
	Learning	0.0211 /seg 0.2 /32seg
	Total	0.0235 /seg 0.2448 /32seg

Table 4.4. *Inference and learning times of the Actor-Critic*

In real-time application, real-time ML consists of generating more accurate predictions and adapting models to changing environments by using real-time data. It outlines the ability to make real-time predictions, ingest new data and update the model accordingly. The ability of real-time prediction is divided into two components:

fast inference and real-time pipeline. For the inference and learning phases, our method performs on a scale of milliseconds. Our pipeline prioritizes stream processing over batch learning and event-driven learning over request-driven learning. To sum up, our approach satisfies the requirements of SA levels 2 and 3 and allows the SA update to perform in real time.

Anomaly detection results

By experimenting, the learning allows us to validate our approach. Figure 4.13 shows that as the episodes progress, when the loss decreases, the agent makes fewer and fewer errors and when the reward increases, the agent ranks video segments better and better during the training phase.

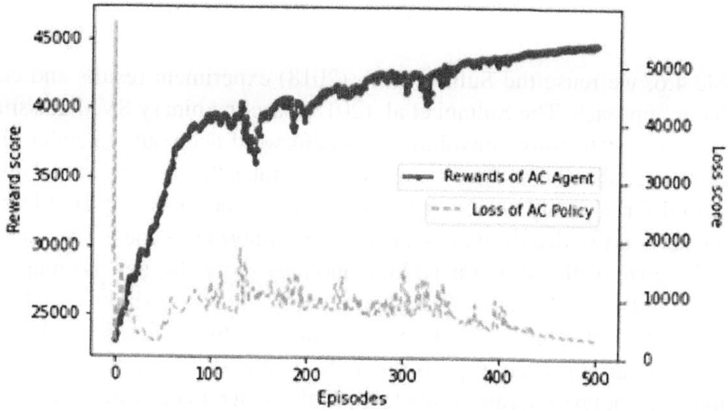

Figure 4.13. *Actor-Critic loss and reward curves*

In Table 4.5, the results show that our method demonstrates its feasibility and verifies that our theory has practical potential. A two-level assessment is provided: the segment level to interpret results and the frame level to compare with existing works.

Detection level	AUC %
Segment-level	77
Frame-level	62

Table 4.5. *Actor-Critic experiment results*

Anomaly detection comparison with the state of the art

To assess our work, we compare our method's performance of 62.66 AUC for frame-level detection with results obtained in scientific work as shown in Table 4.6.

Method	AUC %
Binary classifier (Sultani et al. 2018)	50.0
Sultani et al. (2018); Hasan et al. (2016)	50.6
Proposed	**62.66**
Sultani et al. (2018); Lu et al. (2013)	65.51
Sultani et al. (2018)	75.41

Table 4.6. *Methods comparison on UCF-Crime with our proposed Actor-Critic method*

In Table 4.6, we reuse the Sultani et al. (2018) experiment results and compare them with our approach. The Sultani et al. (2018) baseline binary SVM classifier and the Hasan et al. (2016) fully convolutional feedforward deep auto-encoder obtained lower performance than ours. The first one demonstrates that the traditional approach cannot be used for anomaly detection in the real world because extracted features in the training set are not discriminative enough for abnormal scenes. The second one does not generalize well and produced high anomaly scores for new normal patterns. The Bowling and Iyer (2019) dictionary approach achieves a slightly higher result than ours, but it is not robust enough to discriminate between normal and abnormal. In contrast with our approach, it performs only with past data, without any forecast information. Our approach can compete with the latter one, with an added value "*V* value forecasting" which was not considered in their approach. The proposed method of Sultani et al. (2018) performs better than ours because it manages anomaly recognition in the temporal axis with their framework and their loss contribution designed for MIL-supervised learning. Our AUC rate can be justified by the number of episodes that the agent went through the data. A total of 500 episodes is considerably lower than the 20K epochs used to train an NN with supervised learning in the approach of Sultani et al. (2018). Our model goes through the full dataset 40 times less during the training process. The reason behind the setting of a low number of episodes is because of the training time in RL, which strongly increases by computing *V* value at each step. In other terms, in order to learn to reduce errors at each step, we need to calculate the critic loss called TD error first and then the actor loss. This sequential computing does not allow parallelization inside one learning loop.

By analyzing segment-level results, our approach performed fairly well but has some limitations due to the nature of the dataset. Dealing with this raw dataset, without cleaning, forces us to proceed with MIL and results in a lower frame-level performance. But in our case, with RL, we need to use accurate information about ground truth and not use the concept of MIL, which allows incorrectly located instances inside the wrong bag. Bad labels used in the learning process resulted in wrongly classified scenes.

The MIL introduced semi-supervised learning in order to reduce the time-consuming data process. However, the instance learning concept causes problems because it gives bad guidance to the agent, which blindly trusts rewards. So, to stay in semi-supervised learning, weak supervision is preferred, thus reducing time-consuming tasks. Hence, we will pursue semi-supervised learning but use it differently.

Anomaly detection comparison with the previously proposed method

Compared to the HMM combined SOM method, it provides a poor performance in terms of accuracy as shown in Table 4.7.

Metric	HMM w/ SOM	Actor-Critic RL
ACC %	82	54
AUC %	65	62

Table 4.7. *Comparison study between our interpretable and adaptive model*

4.4.3. *Result-oriented classifier study*

To answer mission-critical performance needs in terms of accuracy, two approaches will be presented, which were not explored in our literature review. The first one is known as a modular approach, which mixes the two approaches presented in capabilities-oriented research; the second is known as the customized learning scheme approach and consists of increasing the performance of the adaptive model in order to compete with the literature in terms of capabilities and accuracy.

4.4.3.1. *Hybrid approach: towards SOM and RL*

The current findings from previous studies have revealed that relying solely on statistical models for interpretable ML and only using adaptive models with RL might not be adequate. Thus, it is suggested that the performance of RL could be enhanced by incorporating relevant features. In order to achieve this enhancement, we have proposed two transition propositions based on our experimental plan. The first proposition is to investigate the incorporation of SOMs to an NN with Multiple Instance Learning (MIL) loss. The second proposition is to explore the incorporation of SOM to an NN with Binary Cross-Entropy loss. Our final proposition is to investigate SOM with an NN-based RL algorithm using its policy-gradient loss.

4.4.3.1.1. Proposed method

This method consists of performing SOM and FFNN with policy-gradient loss. Here, SOM clearly shows its importance when using a distribution-based objective function modeling without using complex MIL loss; the experiment led to the use of it with an RL algorithm as in Figure 4.14. For our application, one of the most basic policy-based approaches is used to check its feasibility, called "REINFORCE algo", which uses the PG objective function.

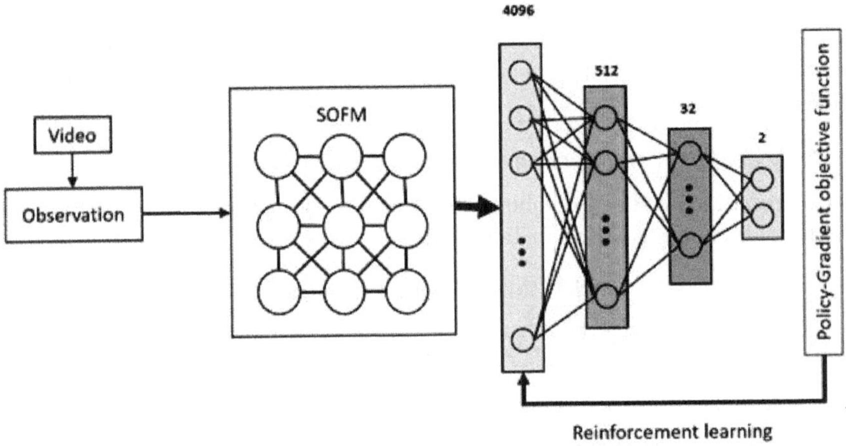

Figure 4.14. *RL and SOM-based anomaly detection framework*

4.4.3.1.2. Experiments

In Figure 4.15, results obtained make it possible to show that SOM could perform well with the RL algorithm, and it performs better than a more sophisticated and better RL algorithm than when not using the SOM, as shown in Table 4.8.

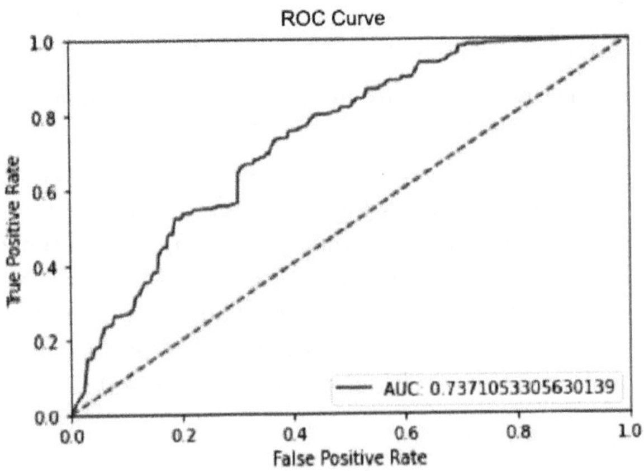

Figure 4.15. *RL and SOM-based anomaly detection ROC curve*

Metric	Actor-Critic w/o SOM	PG w/ SOM
AUC %	65	73
ACC %	54	69

Table 4.8. *RL algorithm comparison study with and without SOM*

These results clearly show a better improvement of the RL algorithm thanks to the SOM implementation even in supervised configuration. The SOM clearly shows its filtering power in this approach, where an MIL approach was initially preferred as a solution when the label is noisy.

4.4.3.2. *New framework and learning scheme: similarity-based meta RL for multi-instance anomaly detection*

In this section, our goal is to develop a customized learning scheme that provides overall control over the internal learning mechanism. Before presenting our novel framework, we first proposed a transition framework called "Joint Classification and Regression via Deep Multi-Task Learning", which surpasses previous methods. However, we are not entirely convinced of its capabilities, as it does not effectively address real-world problems.

In this section, we will focus entirely on our novel framework, "Similarity-Based Meta RL for Multi-Instance Anomaly Detection", which is explained in Figure 4.16.

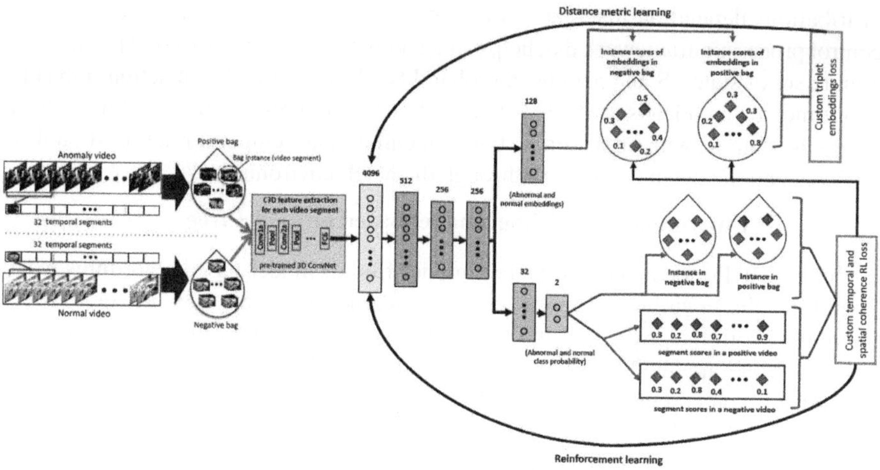

Figure 4.16. *Flow diagram of the proposed anomaly detection and anticipation approach using meta reinforcement learning*

To show off our proposal as in Figure 4.16, selected ML approaches are described, and customized loss/objective functions and learning strategy are explained. This proposal was designed to address all the problems stated in section 4.3.

4.4.3.2.1. Proposed method

A) **ML approaches – similarity-based meta-RL**

Here, we will present the two main approaches used in our solution with their selection criteria. Combining RL and Similarity Learning (SL), our approach, called the "similarity-based meta-RL" method, will permit multi-instance anomaly detection and anticipation. First, we will introduce RL and SL as ML methods.

(a) RL

To understand this approach in depth, we highlight some selection criteria (Sutton and Barto 2005; Li 2017) to understand the future choice.

i. **Environment modeling: temporal extension or not**

First, the environment modeling should be defined as an MDP environment or Multi-Armed Bandit (MAB). Bandit problems are stateless. Each arm has a fixed distribution of rewards. It does not depend on which arms were pulled previously. The goal is to explore the reward distributions of all arms and then keep pulling the best one. MDPs are a temporal extension of bandit problems: pulling an arm influences future rewards. Technically, there is a state that changes by pulling an arm. The reward distributions depend on that state. An MDP is defined as a discrete-time stochastic control process: a framework that helps an agent make a choice. An MDP is built with a finite set of states S and actions, a probability $P_a(s, s^{'})$ to take an action a in state s (at time t) which leads to state s' (at time $t + 1$) and a reward $R_a(s, s^{'})$ to go from state s to state $s^{'}$ with an action a. For sequential data, temporal relation should be modeled and the agent could learn through the MDP environment.

ii. **Environment modeling: model-based or model-free**

In the MDP-modeled environment, model-free or model-based algorithms exist. Therefore, the environment model could be learned to create a policy or to learn the policy from the environment while adhering to MDP properties. RL algorithms based on MDP are divided into two parts: model-based and model-free. Model-based methods rely on certain assumptions, which can sometimes lead to incorrect results. Indeed, if the model is inaccurate, the risk is learning with something far from the original environment. In contrast, model-free methods are very close to reality. They will act and learn from the result. In model-free environment modeling, both value-based and policy-based methods exist, allowing for learning from the environment according to MDP properties.

iii. **Reward availability: dense or sparse reward**

Sparse and dense rewards are two types of reward structures in RL. Sparse rewards are rewards that are given infrequently. In other words, an agent only receives a reward upon completing a specific task or achieving a specific goal. Sparse rewards can often lead to slow learning and difficulty in training an agent. On the contrary, dense rewards are rewards that are given frequently, even for small steps taken towards the goal. Dense rewards can help speed up learning and make it easier to train an agent. Unfortunately, in real-world scenarios, it is more complicated to provide rewards.

iv. **Policy learning: policy-based or value-based**

Policy-based learning is the optimization of the policy via state-value function until it has the optimal one, whereas value-based learning consists of obtaining the best action-value function and then calculating the associated policy.

v. **Policy modeling: shallow or deep architecture**

In deep RL, the policy is adjusted approximately using NNs. NNs are used in addition to standard RL when the latter alone cannot deal effectively with high and complex observation space, which could be represented in CCTV by two- or three-dimensional data.

vi. **Exploration/exploitation dilemma: uncertainty modeling and exploration function**

The chosen exploration function is the Boltzmann exploration function, which is a simple softmax function that will model the probability of each possible action; if the probability of abnormal and normal is tight, that shows that the uncertainty is here, so it needs to explore more to reduce this uncertainty. By sampling with the distribution of the prediction, it will choose if it will select the action opposed to the predictor, i.e. exploration, or if it will follow the prediction of the model, i.e. exploitation. With this approach, we can have a clear model uncertainty regarding each data point. It will logically reduce the uncertainty as it explores and, in the beginning, it will need human input that will be progressively reduced until it converges, and thereafter, humans will be requested only if needed, i.e. when uncertainty appears for certain data points. This function will allow it to learn how to predict and have control over unknown and often stochastic environments when it has never seen data, for instance. In an evolving context, it will deal with the cost of labeling linked to the active learning process. Human guidance will be requested only when needed.

vii. **Higher level policy: meta-RL or not**

Meta-RL is a subfield of RL that involves improving the performance of an agent in RL tasks by learning how to learn. Meta-RL involves training an agent to learn how to quickly and effectively adapt to new tasks and environments by generalizing from experience. In Meta-RL, the agent is trained on multiple related tasks that share some common features, and it then uses this knowledge to quickly adapt to new tasks through a process called "meta-learning". This involves generating a meta-policy, or a higher-level policy, that guides the agent's behavior during a task. Meta-RL focuses on designing algorithms and architectures that facilitate the discovery of effective policies that can learn from small amounts of data and continually improve over time, allowing for faster and more efficient learning in new environments.

- SL

This second approach will enhance the first one by recognizing spatial patterns in higher-level encoding. It will use a distance metric over embeddings. We will first introduce distance metric learning and then embedding learning.

i. **Distance metric learning**

Distance metric learning is a subfield of ML that uses statistical techniques to learn a distance metric between data points in a dataset. The learned distance metric is used to measure the similarity or dissimilarity between data points, and this can be used for tasks such as classification, clustering or retrieval. The goal of distance metric learning is to learn a metric that reflects the underlying structure of the data, such that similar data points are assigned smaller distances in the metric space. Distance metric learning methods generally use a training set of labeled or unlabeled data to learn the metric. The learned distance metric can be either Euclidean or non-Euclidean, and there are various algorithms that can be used to learn the metric. Distance metric learning has important applications in a range of fields such as computer vision, natural language processing and recommender systems. It plays a crucial role in feature extraction, similarity matching, and information retrieval tasks and has the potential to significantly improve the performance of ML systems.

ii. **Embedding representation learning**

An embedding is a relatively low-dimensional space into which we can translate high-dimensional vectors. Ideally, an embedding captures some of the semantics of the input by placing semantically similar inputs close together in the embedding space. An embedding can be learned and reused across models. Embedding learning consists of proceeding representation learning through distance metric learning as a triplet or contrastive learning, for instance. Representation Learning is concerned with training ML algorithms to discover useful representations, for example those that are interpretable, have latent features, or can be used for transfer learning.

For the first part, the aim of an RL task is to optimize the objective function O_f with the objective of maximizing the expected cumulative reward (equation [4.4]):

$$O_{f_1} = \max_{\theta} \mathbb{E}_{\pi_\theta} R(\tau) \quad [4.4]$$

where $\pi(\theta)$ is a parametrized policy (typically an NN) and τ is a trajectory which is a sequence of states and actions. The expected value is taken with respect to drawing trajectories τ under the policy $\pi(\theta)$ and so solving this problem is equivalent to finding the "best" parameters θ that give the best policy for maximizing the expected reward.

For the second part, anomaly detection is considered as a regression problem. It looks for anomalous video segments with higher anomaly scores than the normal segments. The straightforward approach would be to use a ranking loss, which encourages high scores for anomalous video segments as compared to normal segments, such as:

$$f(\mathcal{V}_p) > f(\mathcal{V}_n) \qquad [4.5]$$

where $\mathcal{V}_p$ and $\mathcal{V}_n$ represent the positive and negative video segments, and $f(\mathcal{V}_p)$ and $f(\mathcal{V}_n)$ represent the corresponding predicted anomaly scores ranging from 0 to 1, respectively. The above ranking function should work well if the segment-level annotations are known during training.

B) **ML methods – objective and loss functions**

(a) The spatial objective function for MIL in RL

In the spatial domain, the multi-instance anomaly detection is handled as an MAB problem, considering states uncorrelated by the time in τ.

The return obtained by the agent will constitute the QoP, the QoP in this approach consists of rewarding the entire trajectory at once as follows:

$$QoP(\tau) = \begin{cases} 1, & \text{if all } f(\mathcal{V}_n) < 0.5 \\ & \text{or } (\max_{i \in \mathcal{B}_p} {}_1 f(\mathcal{V}_p) >= 0.5 \\ & \text{and } \max_{i \neq \max_1, i \in \mathcal{B}_p} {}_2 f(\mathcal{V}_p) >= 0.5) \\ 0, & \text{otherwise} \end{cases} \qquad [4.6]$$

All the segments in the negative bag should be under the half, and at least the max two segments in the positive bags should be over the half. In MIL, at least one segment in the positive bag is most likely to be positive, but as we want to integrate a temporal coherence, we will push and correlate one segment to another one, and for that, this coherence should happen at least between two segments.

In the absence of state correlation, the spatial MI-RL optimization will occur with equation [4.4], where $\pi(\theta)$ is a parametrized policy and R is obtained by equations [3.1] and [3.2], where ${}_{trad}QoE$ and ${}_{aug}QoE$ are equal to a null value. The expected value is taken with respect to drawing trajectories τ under the policy $\pi(\theta)$ and so solving this problem is equivalent to finding the "best" parameters θ that give the best policy for maximizing the reward R.

(b) The spatio-temporal objective function for MIL in RL

In the temporal domain, we handle the temporal anomaly detection as MDP considering states correlated by the time in τ.

Our proposal considers the temporal coherence of sequences via MDP with RL in addition to spatial coherence done via MAB with RL.

Looking at temporal and spatial coherence is relevant to the loss function. For this, we will investigate the objective function deeply.

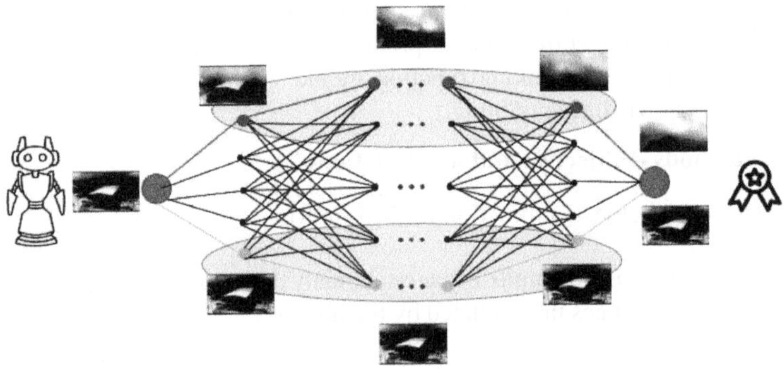

Figure 4.17. *N-step reinforcement learning in an MDP environment*

As in Figure 4.17, behavior is modeled as a trajectory, abnormal trajectories and normal trajectories. The goal of a predictor is to generalize well around these two ideal behaviors, abnormal and normal. Overfitting will mean that the predictor is predicting existing trajectories too accurately, and underfitting will mean that it will not represent enough trajectories.

Here, the QoP built on the reward process is represented as follows:

$$QoP(s_t, a_t) = \begin{cases} 1, & \text{if } f\left(\mathcal{B}_n\right) < 0.5 \\ & \text{or } f\left(\mathcal{B}_p\right) >= 0.5 \\ 0, & \text{otherwise} \end{cases} \qquad [4.7]$$

Temporal utility analysis justifies the evaluation of options according to their futures. It was possible thanks to the Bellman equation, which gives us the ability to describe the value of the state s_t, $V(s_t)$, with the value of the state s_{t+1}, $V(s_{t+1})$

and with an iterative approach, we can calculate the values of all states. To build our $TemporalUtility$, we compute recursively the discounted reward as follows:

$$TemporalUtility(\tau) = \sum_{t=0}^{T-1} \gamma^t R(s_t, a_t) \quad [4.8]$$

where $\gamma \in [0, 1]$ is the discount factor that varies the vision depth of the agent for anticipation, R is the reward, and t is the discrete time. Define $R(s_t, a_t)$ to be the consumer preference received after observing the state s_t and performing an action a_t. Also, define the discounted sum of these rewards as $TemporalUtility(\tau)$, which is the temporal utility containing time-augmented situational information. It considers states as weighted and allows for anticipation based on a discount factor that determines how deep the agent can anticipate.

$$O_{f2} = \max_{\theta} \mathbb{E}_{\pi_\theta} TemporalUtility(\tau) \quad [4.9]$$

where $\pi(\theta)$ is a parametrized policy (typically an NN). The expected value is taken with respect to drawing trajectories τ under the policy $\pi(\theta)$ and so solving this problem is equivalent to finding the "best" parameters θ that give the best policy for maximizing the $TemporalUtility$.

In the spatio-temporal domain, the multi-temporal instance anomaly detection is handled as both MAB and MDP problems.

A strong temporal coherence can act like a sequence learner and classify the set of all sequences as a big sequence and classify them by taking into account only the spatial coherence. The objective is to balance temporal and spatial coherence effectively to perform MIL by appropriately considering the relationships in both temporal and spatial domains during the classification task.

Our Spatio-Temporal Multi-Instance RL objective function is therefore given as follows:

$$O_{f3} = O_{f1} + O_{f2} \quad [4.10]$$

This custom objective function embedded a spatio-temporal coherence between instances.

(c) The multi-instance triplet loss for spatial pattern recognition

To enhance our approach, our model will be regularized by learning task-invariant features for seen and unseen tasks via similarity-based meta-learning. When RL suffers from catastrophic forgetting, one solution adopted is called latent learning, which is a technique used to mitigate catastrophic interference by taking advantage of transfer learning. This approach tries to find optimal encodings for any new classes to be learned so that they are less likely to catastrophically interfere with existing responses. In our approach, this latent encoding will act as long, reliable and compressed memory, permitting fast learning and reliable regularization.

However, in the absence of video segment-level annotations, it is not possible to use a ranking loss that encourages high scores for all anomalous video segments, as compared to all normal segments. Instead, we propose to use the following multiple instance embeddings ranking objective function:

$$\max_{i\in\mathcal{B}_p} E\left(\mathcal{V}_p^i\right) > \max_{i\in\mathcal{B}_n} E\left(\mathcal{V}_n^i\right) \qquad [4.11]$$

where max is taken over all video segments in each bag.

In the absence of scores, we retrieve the max embedding through the arguments of the maxima scores obtained by a score predictor as in the following multiple instance embedding ranking objective function:

$$E\left(\arg\max_{i\in\mathcal{B}_p} f\left(\mathcal{V}_p^i\right)\right) > E\left(\arg\max_{i\in\mathcal{B}_n} f\left(\mathcal{V}_n^i\right)\right) \qquad [4.12]$$

The segment corresponding to the highest anomaly score in the positive bag is most likely to be the true positive instance (anomalous segment). The segment corresponding to the highest anomaly score in the negative bag is the one that looks most similar to an anomalous segment but actually is a normal instance.

This negative instance is considered a hard instance, which may generate a false alarm in anomaly detection. By using equation [4.13], as in Figure 4.18, we want to push the highest anomaly score positive instances and highest/all negative instances far apart and bring closer the highest anomaly score negative instances and highest/all negative instances in terms of embeddings.

Our ranking similarity loss in the triplet formulation applied to embeddings is therefore given as follows:

$$\mathcal{L}(A, P, N) = \max\left(\|A - P\|^2 - \|A - N\|^2 + 1, 0\right) \qquad [4.13]$$

$$\text{with} \begin{cases} A : E\left(\mathcal{V}_n\right) \text{ or } E\left(\arg\min_{i\in\mathcal{B}_n} f\left(\mathcal{V}_n^i\right)\right) \\ P : E\left(\arg\max_{i\in\mathcal{B}_n} f\left(\mathcal{V}_n^i\right)\right) \\ N : E\left(\arg\max_{i\in\mathcal{B}_p} f\left(\mathcal{V}_p^i\right)\right) \end{cases}$$

where A is an anchor input, P is a positive input of the same class as A, and N is a negative input of a different class from A.

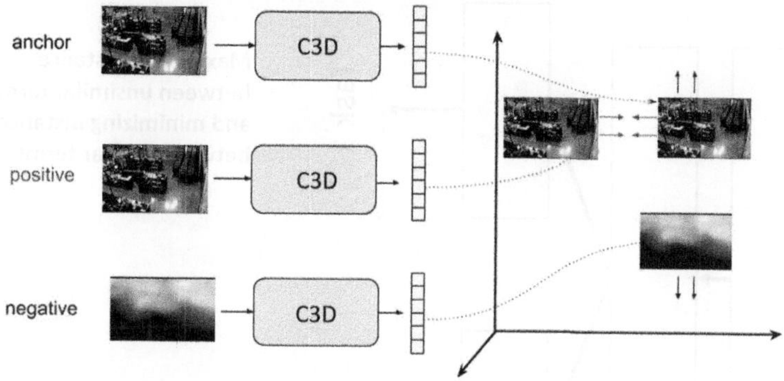

Figure 4.18. *Distance metric learning with triplet loss*

C) ML learning strategy – concurrent and cooperative learning

To optimize these two objective functions, one for the SL and the other for the RL, a suitable tailored training strategy should be selected and developed. For this, these multi-objective function optimizations could be seen as a multitask learning problem. Multi-task learning (MTL) is a subfield of ML in which multiple learning tasks are solved at the same time, while exploiting commonalities and differences across tasks. This can lead to improved learning efficiency and prediction accuracy for the task-specific model, compared to separate training models.

MTL works because regularization induced by requiring an algorithm to perform well on a related task can be superior to regularization that prevents overfitting by penalizing all complexity uniformly as dropout or regularization terms. Within the MTL paradigm, information can be shared across some or all of the tasks. Depending on the structure of task relatedness, we may want to share information selectively across the tasks. Traditional MTL implies that a shared representation is developed concurrently across tasks, also known as concurrent learning. Information sharing occurs at both low and high levels. In our case, this means that the low-level transfer takes place in the shared layers, and the high-level transfer occurs at the prediction level when providing abnormal scores from RL predictor to compute max embeddings.

As in Figure 4.19, task 2 will optimize the RL loss to predict the probability that the state belongs to the normal or abnormal class. Task 1 is to maximize the distance between non-similar embeddings and minimize the distance between similar embeddings.

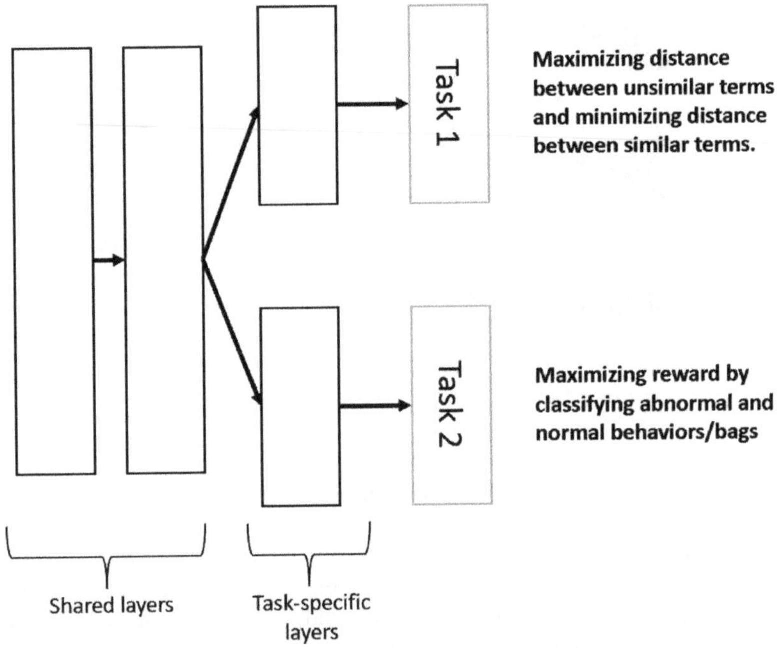

Figure 4.19. *Multi-task learning: reinforcement and similarity learning*

The training will be as follows:

– optimize RL loss video after video to predict the probability that the state belongs to the normal or abnormal class;

– once the number of stored videos is reached, optimize the similarity loss on the batch to maximize the distance between non-similar embeddings and minimize the distance between similar embeddings.

The number of states containing the video and the length of the batch are parameters to be set. The custom algorithm is inspired by the most generic algorithm for policy-based methods in RL, called the REINFORCE algorithm (Williams 1992). We adjust the new RL loss and provide a memory to build a batch on what was learned for SL. The proposed method, custom loss and personalized training strategy are defined as similarity-based meta-RL.

4.4.3.2.2. Experiments

By experimenting, the learning allows us to validate our approach. As shown in Figure 4.20, as the episodes progress, the agent makes fewer and fewer errors and when the reward increases, the agent ranks video segments better and better during the training phase. The learning was stopped by the early stop rules defined for our experimentation in episode 51. Knowing that the size of our batch is set to 60 and that there are a total of 1,920 videos, this means that we need 32 batches to go through the entire dataset.

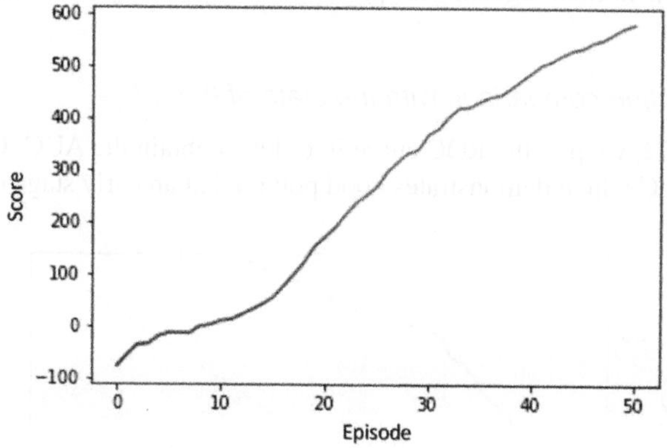

Figure 4.20. *Learning curve with average cumulative rewards*

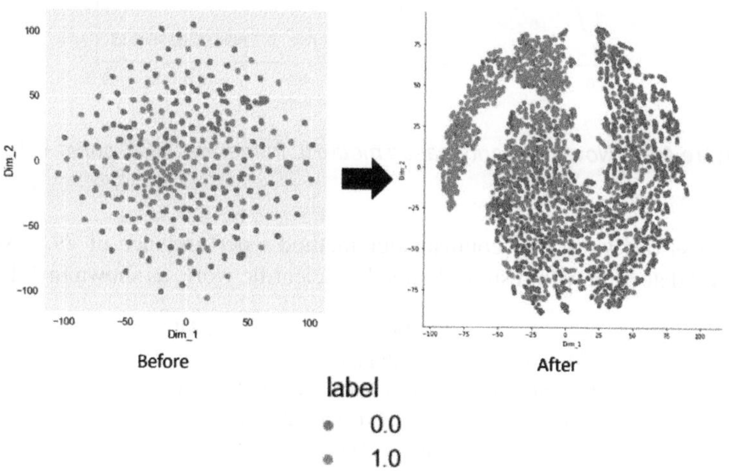

Figure 4.21. *2D visualization of embeddings on testing data with t-SNE before and after applying distance metric learning*

In Figure 4.21, we plot on the left features reduced to 2D by the TSNE algorithm (van der Maaten and Hinton 2008), which is a dimension reduction algorithm that is used here to show in 2D the spatial representation of data. On the right, we plot the embeddings after the 51 episodes of learning. After learning, in the right plot, we can see that it leads to a concentration of the information of the abnormal scene in orange on the left and that the information of the normal scene is concentrated on the right. The points of the abnormal and normal scenes are intermingled because we plot information that, in some cases, is inaccurate. This is because, in the video-level annotations, some normal segments are present in the abnormal videos and are labeled as abnormal.

Anomaly detection comparison with the state of the art

In Figure 4.22, we plot the ROC curve in order to obtain the AUC. Our method is rated at 79.9 AUC which demonstrates good potential at an early stage of learning.

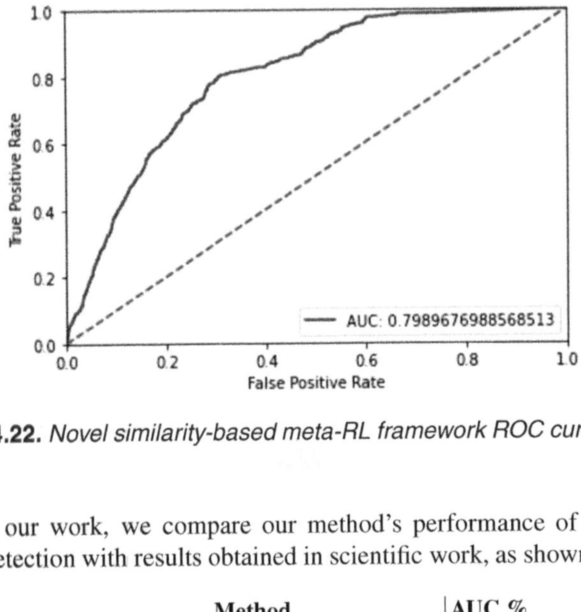

Figure 4.22. *Novel similarity-based meta-RL framework ROC curve – AUC*

To assess our work, we compare our method's performance of 79.9 AUC for frame-level detection with results obtained in scientific work, as shown in Table 4.9.

Method	AUC %
Binary classifier (Sultani et al. 2018)	50.0
Sultani et al. (2018); Hasan et al. (2016)	50.6
Sultani et al. (2018); Lu et al. (2013)	65.5
Li et al. (2021)	72.9
Sultani et al. (2018)	75.4
Aberkane and Elarbi (2019)	78.2
Proposed	**79.9**
Lv et al. (2020)	85.4

Table 4.9. *Methods comparison on UCF-Crime with our similarity-based meta-RL method*

In Table 4.9, we reuse experiment results from Sultani et al. (2018) and compare them with our approach. The baseline binary SVM classifier in Sultani et al. and the fully convolutional feedforward deep auto-encoder in Hasan et al. (2016) obtained lower performance than ours.

The first one demonstrates that the traditional approach cannot be used for anomaly detection in the real world because extracted features in training sets are not discriminative enough for abnormal scenes. The second one does not generalize well and produces high anomaly scores for new normal patterns. Both of them obtained scores relatively near to 50, which is close to a random predictor.

Lu et al. (2013) dictionary approach and generative and clustering approach in Li et al. (2021) get a slightly higher result than the first ones but are not robust enough to discriminate between normal and abnormal. The generative and clustering approaches get lower results compared to ours because they use the generative model, which has a high generalization capacity that could encode normal as abnormal ones, and the clustering does not guarantee the separation of abnormal and normal and does not guarantee the provision of good annotations to the generative model.

The method proposed by Sultani et al. (2018) performs better than previous ones because it manages anomaly recognition well on the temporal axis with their framework and their loss contribution is designed for MIL-supervised learning. It obtains lower results than ours because it does not take into account a temporal coherence as strong as the one we consider and because we reworked the C3D features to obtain more representative embeddings. Aberkane et al. (2019) reuse segment-level annotation provided by Sultani et al. to proceed to RL. These results are a little better than the Sultani et al. approach because it will generalize over the segment-level information. Their approach uses RL as a standard classification algorithm. The proposed method by Lv et al. (2020) performs much better than previous methods with an AUC score of 85.4, which is the third-highest score in state of the art (SoTA); this score is closely linked to feature extraction methods. His study shows that using only semantics yields a score of 82%. However, adding dynamic variation does not significantly improve this score. The real impact appears when training on the augmented dataset with noises and handcrafted anomalies.

In contrast with our approach, although other approaches perform a temporal analysis taking into account the surrounding information, i.e. the past and the future, they do not provide forecast information on the impact of these states in relation to an abnormal event in order to anticipate it. Our approach competes with them, with an additional "V value" that considers states as weighted and which is not taken into account in the other approaches. This is the reason why it takes time to compute, but it provides information on the normality or abnormality of future states compared to current states. Another limitation is that we compare the ground truth provided by Sultani et al. with SoTA, but in our approach, we do not only consider the spatial anomaly called temporally localized anomaly, but also the temporal coherence to predict in advance when an anomaly will occur for anticipation. Ultimately, ground truth is not appropriate if the few segments preceding an anomaly are considered as an error.

In Table 4.10, our proposed methods have been measured by important metrics (ACC, AUC, FN, TN) selected for our use case with critical properties. Our first proposed methods consider only the spatial domain and perform multi-instance learning with RL, which leads to a good accuracy of 83 and an AUC of 77, which is not bad. To this, we decided to add a temporal extension, which returns a result of 80 in ACC, which is our limit not to be exceeded, and a result of 79.4, which is more interesting because the FN rate decreases and the TP increases compared to the former. To improve the results, we incorporate SL on the embeddings to have a better data representation space, and this provides a score of 80 in precision and 79.9 in AUC with a better TP rate and a decreased FN rate. Between the first and the second method, the additional task of taking into account the temporal extension plays the role of regularization. Moreover, the addition of the embedding approach makes it possible to divide the learning time by four as well. This "similarity-based meta-RL for multi-instance anomaly detection" approach proposed the best results among our proposals.

Method	ACC %	AUC %	FN %	TP %	NB. of batch
Proposed	**83**	77	4.28	3.31	413
Proposed w temporal coherence	80	79.4	3.6	3.9	211
Proposed w temporal coherence & embeddings	80	**79.8**	**3.1**	**4.4**	**51**

Table 4.10. *Ablation study on proposed methods*

4.5. Conclusion

This chapter discusses a new predictive system with situational awareness as its focus. The goal is to enhance the system's abilities to meet mission-critical requirements and provide satisfactory results for anomaly detection in real-world situations. To achieve this, real-world problems in video surveillance, such as temporal anomaly localization, concept drift, model interpretability and trail error learning, were addressed. Our research was divided into two phases: one to improve our model's capabilities and the other to increase its performance rate.

In the first phase, we explored popular modular approaches and proposed a self-competitive NN SOM with an HMM to make multi-instance learning interpretable. We also explored the Actor-Critic RL algorithm with a neural network to make continuous learning-based decisions.

In the second phase, we focused on improving performance by being aware of the problems we needed to address. To achieve this, we first proposed to combine the previous two methods and suggested using an SOM and NN-based RL algorithm to detect abnormal situations. This approach provided results of 73% AUC, which is satisfactory but not sufficiently competitive, even though it tackles other challenges not addressed in the literature. Finally, we proposed a learning scheme and offered a "similarity-based meta-RL for multiple instances anomaly detection", which allowed us to address all stated problems and achieve competitive results of 79.9% AUC.

In conclusion, this last approach enables our predictive system to ensure situational awareness for abnormal situation detection, even in unfavorable conditions.

5

Towards an Autonomic Intelligent Video Surveillance System

"The key with autonomous is the whole ecosystem. One of the keys to having truly fully autonomous is vehicles talking to each other."
Mary Barra

Chapter 5 presents a novel autonomic approach for an Intelligent Video Surveillance (IVS) system. The purpose is to let the system self-manage to gain ownership and performance by reducing user management time.

5.1. Context

When companies aim to bring their products and processes to market, they must conduct comprehensive research and development to ensure readiness for entry.

To ensure successful product and process development, realistic testing must be conducted and demonstrated under practical conditions. It is important to have access to the appropriate support infrastructure and equipment, depending on the specific development phase.

The process of scaling up a prototype developed during the experimentation phase and transferring it to meet production expectations is known as the scale-up process. This involves transferring a manufacturing process or section of a process from the laboratory scale to the level of commercial production.

For a color version of all the figures in this chapter, see www.iste.co.uk/djeachandrane/smart.zip.

However, one aspect that has been largely overlooked is the transition of innovation from a small-scale prototype to larger-scale production. This change in scale is a crucial consideration in the development and implementation of new technology.

It raises important questions. After demonstrating its worth in a particular area:

– How can we ensure that it has benefits in other areas as well?

– How can we spread it, replicate it in different regions or apply it to other sectors?

– What is the role of the stakeholders, vendors and customers involved in the solution?

The goal of scaling up is to disseminate an innovation by simplifying its understanding and ensuring that it is easily accessible to everyone involved in implementing and using it in new areas and territories.

5.2. Problem statement

The success of a product's diffusion and generalization depends on whether users find it acceptable (Shackel and Richardson 1991), regardless of the government's economic and political aspects. User acceptability is determined by four factors: utility, usability, likeability and cost. Likeability refers to a user's attitude towards a product and is influenced by its usability and utility. However, for this work, likeability will be indirectly studied through usability and utility.

To examine these factors, it is helpful to consider a practical scenario where assumptions about a system's parameters are relaxed during the production phase. This often happens when there is a lack of pre-existing knowledge or information about the system. The third-party technician may not be aware of the expected configuration, or the configuration may be too complicated to model accurately. In addition, the final user's configuration may not be effectively communicated to the technician, resulting in misinformed assumptions about the system.

Cost includes all expenses related to the process, including maintenance, remediation and configuration. Longer configuration or remediation times, as well as the mobilization of additional task force members, can increase costs. Utilizability refers to a product's practical and effective ability in real-world situations. In the scenario we discussed earlier, it is important to ensure that the product does not increase human workload to the point where it becomes time-consuming or labor-intensive. Utility refers to a product's ability to meet users' expectations and required functionality. In the scenario we discussed earlier, the product's utility will vary from one user to another, and even between the technician who created it and the final user who will be using it. Determining user acceptability involves analyzing the balance between the best possible alternatives of its factors.

In order to enhance user experience, it is advisable to minimize the need for human intervention, which can lead to cost savings and improved usability. However, despite reducing human involvement, it is important to ensure that responsiveness requirements are met. In addition, having greater control over the product or service's usefulness can help maintain ownership and enhance its overall acceptability. These reflections will lead us to the following questions:

– How can management tasks be automated to reduce human intervention?

– How can system responsiveness to management be increased?

– How can enough control be provided over the solution to reduce ownership?

5.3. Proposals

In order to address the problem of user acceptability, we propose developing a solution that will reduce user management time and make the system more user-friendly and increase the responsiveness of the system to its management. Our proposal involves using self-managed autonomic computing. We will develop control mechanisms from scratch, using control theory. A control system is a system that is designed to manage, command, direct or regulate the behavior of other systems or devices. The main objective of a control system is to maintain a desired output in response to changes in input or disturbances. There are various types of control systems, including open-loop, closed-loop and feedback control systems. We will investigate two types of controlled systems: a time-based control system and an event-based control system. A time-based control system is a type of control system that relies on a fixed schedule or time interval to initiate a specific action, and an event-based control system, on the contrary, is a type of control system that initiates a control action based on a specific event or change in a condition or status.

Before presenting our proposed systems, we will first define the dataset and feature extractor, as these are essential elements of our intelligent video surveillance (IVS) system. We propose three systems: the first will allow for self-configuration, the second will enable self-protection and the third will be self-aware and capable of learning.

A) **Dataset**

The dataset (Gamaleldin et al. n.d.) was created by a team during the NASA Space Apps Challenge in 2018, the goal was to use the dataset to develop a model that can recognize the images with fire.

Data is divided into two folders, the fire_images folder contains 755 outdoor-fire images, some of them containing heavy smoke, the other one is non-fire_images which contains 244 nature images (e.g. forest, tree, grass, river, people, foggy forest, lake, animal, road and waterfall). Images get a shape of 224, 224, 3.

B) **Feature extractor**

The feature extractor is designed using a CNN architecture with a convolution and pool layer. This architecture is chosen to extract important data from 2D image data. CNNs are well-suited for image data because they can effectively learn patterns through a filter and aggregation process, which allows them to capture spatial information.

5.3.1. *Time-based control*

5.3.1.1. *User-centric self-configuring decision support system*

5.3.1.1.1. Proposed system

The proposed system in Figure 5.1 is a control system based on time. It will be integrated into a decision support system (DSS) ecosystem and will replace the DSS software component. The control system consists of a prediction function that serves as the controlled system, a learning mechanism that functions as a controller and a feedback generator in place of a signal comparator. In our case, the signal comparator and feedback generator can be interchangeable due to the binary decision.

The unique feature of this approach is the feedback generator, which measures user satisfaction and produces rewards. The goal of this approach, known as "User-centric self-configuring DSS", is to self-configure the system and save the user configuration time during installation.

5.3.1.1.2. Proposed method

To conduct experiments, we augmented the dataset with salt and pepper noise by a factor of 0.11. This resulted in an augmented dataset of 972 images, including the original dataset of 486 images. For the training phase, we split the full augmented dataset into training, validation and testing sets. This allowed us to observe the learning capacity through the learning curve and assess the model's performance on unseen data. We set the training ratio to 0.75, the validation ratio to 0.15, and the test ratio to 0.10. We used a learning rate of 0.001 and trained the model for 100 epochs. We used stochastic gradient descent as the optimizer with a momentum of 0.9 and a decay to the initial learning rate divided by the number of epochs (0.001/100). The loss function used was binary cross-entropy, which is suitable for anomaly detection tasks. Finally, the classifier neural network architecture was configured with dense layers and dropout layers for regularization assignment.

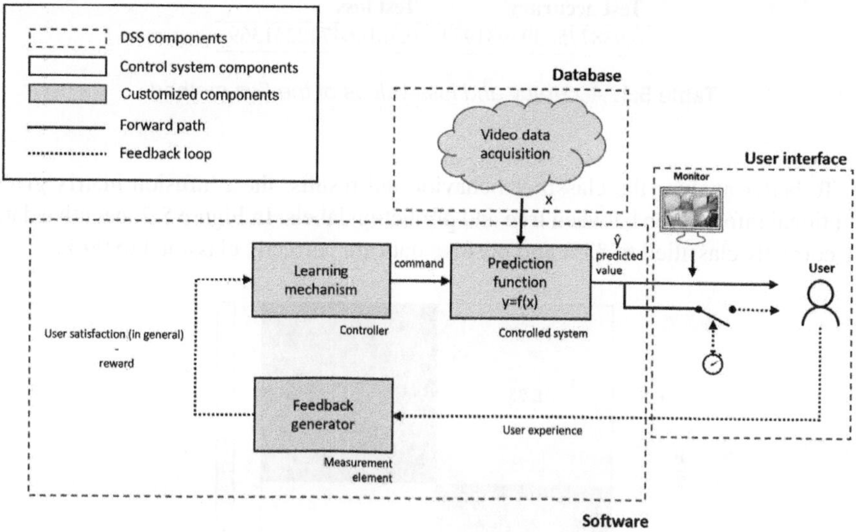

Figure 5.1. *User-oriented self-configuring IVS*

5.3.1.1.3. Experiments

In Figure 5.2, on the left side, the learning curves demonstrate successful learning by consistently increasing and approaching convergence between training and validation accuracies. In Figure 5.2, on the right side, the loss curves demonstrate successful learning by consistently decreasing and approaching convergence between training and validation losses.

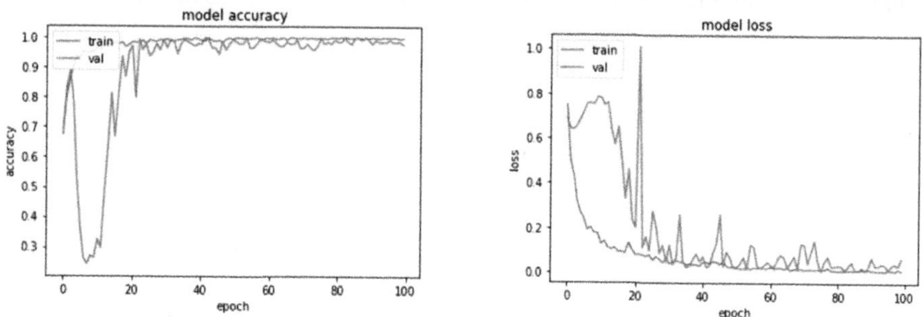

(a) Accuracy curve (b) Loss curve

Figure 5.2. *Accuracy and loss curves of the first method*

To support the results, the classifier was assessed on the testing set, and it provides satisfying results in Table 5.1.

Test accuracy	Test loss
0.9387755393981934	0.361664742231369

Table 5.1. *Accuracy and loss values of the first method*

To better analyze the classifier behavior and results, the confusion matrix gives additional information between true and prediction labels. In Figure 5.3, negative data are correctly classified to 83% and positive data are correctly classified to 99%.

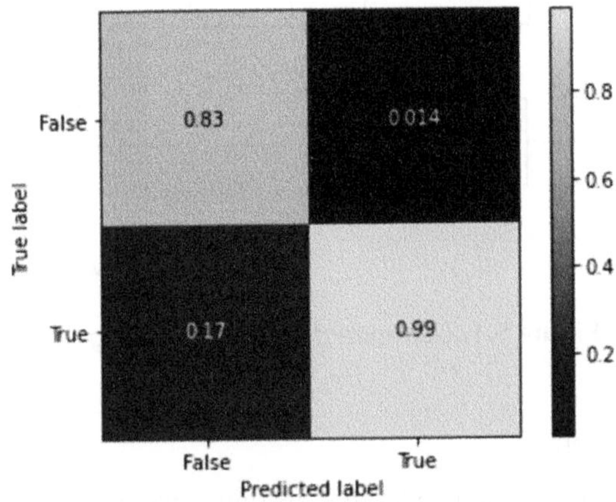

Figure 5.3. *Confusion matrix of the first method*

Once learning proof is shown and the classifier assessment is done, our study is completed by assessing real data and its output on some samples as in Table 5.2.

5.3.2. *Event-triggered control*

5.3.2.1. *User-centric self-protecting DSS*

5.3.2.1.1. Proposed system

In Figure 5.4, the proposed system is a control system which is triggered by an event. As in the previous method, the control system will occupy the place of the DSS software component. This control system is composed of a prediction function which is the controlled system, a learning mechanism as a controller and a feedback generator as a measurement element. The specificity of this approach is first in the feedback generator, which is able to measure the user satisfaction of one preference – analyzed content and second, in the event-triggering mechanism which is distinct from the prediction function and based on the VQA assessment method and statistical tests to trigger an abnormal VQA which is generally when video data quality is deteriorating. This "User-centric self-protecting DSS" approach aims to self-protect the system from intentional or unintentional data source deterioration.

Normal content and good quality	Normal content and bad quality

Real label: 0 Model outputs: [1.00e+00 2.06e-12] Model classification: 0	Real label: 1 Model outputs: [4.49e-05 9.99e-01] Model classification: 1
Abnormal content and good quality	**Abnormal content and bad quality**

Real label: 1 Model outputs: [2.35e-05 9.99e-01] Model classification: 1	Real label: 1 Model outputs: [4.99e-06 9.99e-01] Model classification: 1

Table 5.2. *Model output on real samples of the first method*

5.3.2.1.2. Proposed method

A) **Descriptive and inferential statistics**

As shown in Figure 5.5, data could be affected by biased data quality impacting the predictive system quality as it is a data-driven solution.

To study the data of data quality, a descriptive statistic is used to have a summary statistic that quantitatively describes and summarizes from a collection of information.

Then, inferential statistics are used from the data to learn more about the population the data sample is intended to represent.

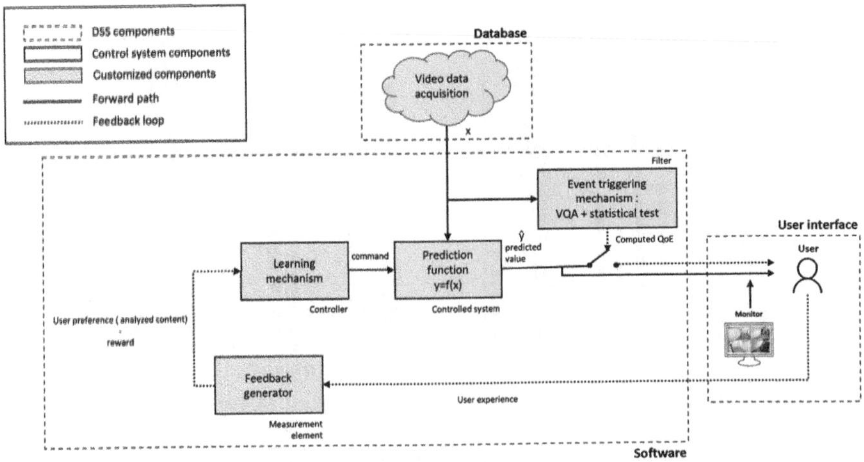

Figure 5.4. *User-centric self-protecting IVS*

Figure 5.5. *Data deterioration due to salt and pepper noise levels ranging from 0.05 to 0.25*

A boxplot is a statistical tool used to display the range, spread and skewness of numerical datasets through their quartiles. It is a descriptive statistics tool that presents the distribution of data in a graphical format. Any outliers that significantly vary from the rest of the dataset can be plotted as individual points beyond the whiskers on the boxplot. Boxplots are non-parametric, meaning they do not depend on assumptions of the underlying statistical distribution. They show the variation in samples of a statistical population without making any assumptions about the data's distribution.

When attempting to draw inferences about a population parameter, which is a characteristic that describes a population, using sample data, hypothesis testing methods such as the Student's t-test, Welch t-test and Mann–Whitney U-test are frequently used.

– The t-test is a popular statistical test used to measure the differences between the means of two groups or a group compared to a standard value. It is based on a probability law called Student's law. The Student's t-test works well when population variances and sample sizes are equal. Welch's t-test is an equivalent test in which data are set with unequal sample variances and/or unequal sample sizes.

– The Mann–Whitney U-test is therefore the non-parametric counterpart of the t-test for independent samples; it is subject to less strict assumptions than the t-test. Therefore, the Mann–Whitney U-test is always used when the normal distribution condition of the t-test is not met.

– In the Mann–Whitney U-test, the test is based on a difference in central tendency, whereas in the case of the t-test, the test is based on a difference in mean values.

In Figure 5.6, boxplots of original data quality and noisy data with 0.05, 0.11 and 0.25 noisy factors are assessed with these three tests empirically.

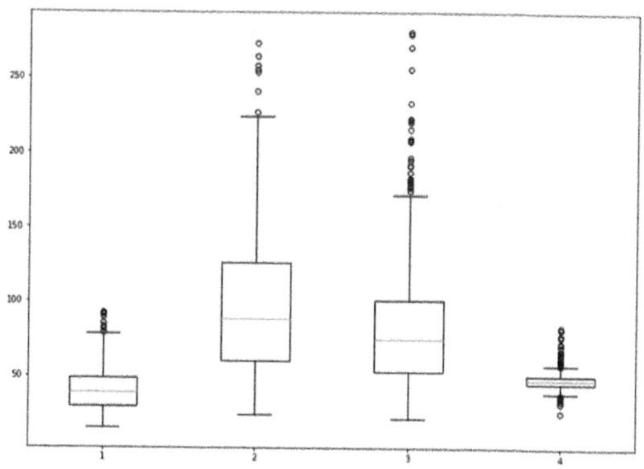

Figure 5.6. *Boxplot comparative study of different levels of noise injected in a population*

The p-value is a value that tells us whether the results of an analysis are significant or not. It is therefore calculated during a statistical test as in Table 5.3. The "P-value" is a statistical measure that helps scientists determine whether their hypotheses are correct.

Salt and pepper noise	Student t-test	Welch t-test	MW U-test
0.05	3.38e-106	1.91e-93	17.3e-123
0.11	1.99e-81	4.63e-74	4.93e-89
0.25	4.03e-28	2.44e-27	7.99e-36

Table 5.3. *P-values of three tests on different levels of noise injected in a population*

In Table 5.3, three tests were conducted to compare the quality of the original data with the quality of data at three different noise levels. In this table, significant results are observed. A statistically significant test result means that the test hypothesis is false or should be generally rejected when the value is less than 0.05, i.e. when it is not in the range 95% of the confidence interval.

Figure 5.8 displays boxplots of the full population of original data and various sample sizes of noisy data: 25, 50, and 100. These datasets will be assessed next using three empirical tests.

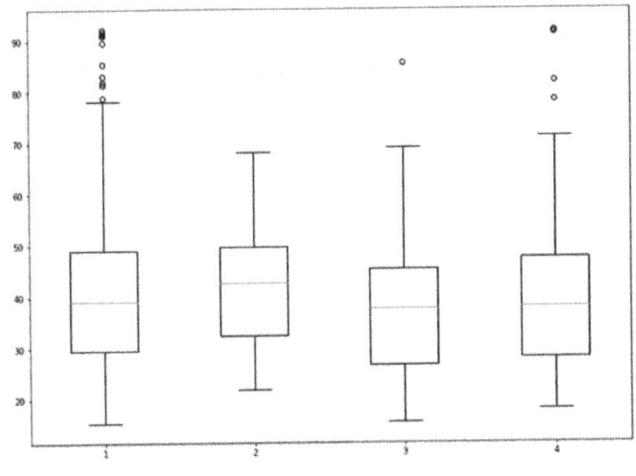

Figure 5.7. *Boxplot comparative study of different sample size of original data*

In Table 5.4, observed results are not significant, which means that the test hypothesis is true, and the hypothesis is accepted. These results show that sample size sampled from the original population does not impact according to this statistical test.

Sample size	Student t-test	Welch t-test	MW U-test
25	0.53	0.43	0.32
50	0.32	0.29	0.08
100	0.70	0.72	0.46

Table 5.4. *P-values of three tests on different sample sizes of original data*

Figure 5.8 displays boxplots of the full population of original data and various sample sizes of noisy data: 25, 50, and 100. These datasets will be assessed next using three empirical tests.

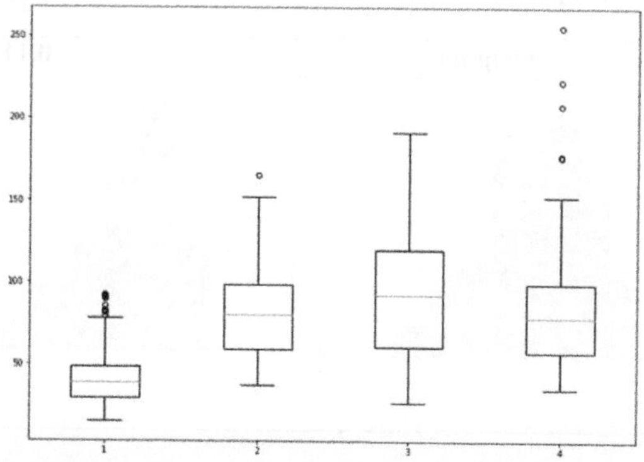

Figure 5.8. *Comparative study of boxplots for the full original dataset and different sample sizes at a data noise of level 0.11*

In Table 5.5, significant results are observed. These results show that samples containing noisy data with different sample sizes are significantly different from the original data without noise.

Sample size	Student t-test	Welch t-test	MW U-test
25	6.65e-34	2.32e-06	1.43e-11
50	5.10e-62	1.23e-11	9.78e-20
100	1.48e-61	2.29e-18	1.03e-34

Table 5.5. *P-values of three tests to compare the full original dataset with different sample sizes of noisy data at level 0.11*

In order to gain a deeper understanding of the data, we first conduct an exploratory data analysis by plotting histograms of the original and noisy datasets with a value of 0.11. These histograms provide us with a visual representation of the distribution of the data and allow us to assess the quality of the datasets. Upon closer inspection, we observe that both datasets exhibit a bell-shaped curve, which is a characteristic of a normal distribution. Our findings are presented in Table 5.6, where we list the key statistical characteristics of the datasets.

To confirm that the data is normally distributed, we decided to fit the datasets with a Gaussian distribution (Gauss 2014). To do this, we employed the least squares method (Gauss 2014) to fit the Gaussian distribution, which is a statistical approach that aims to minimize the sum of the squares of the residuals (the differences between the observed values and the fitted values). The results of this process provide us with a more detailed understanding of the data and enable us to draw more accurate conclusions about its properties.

noise factor	**Original**	**0.11**
Data histograms	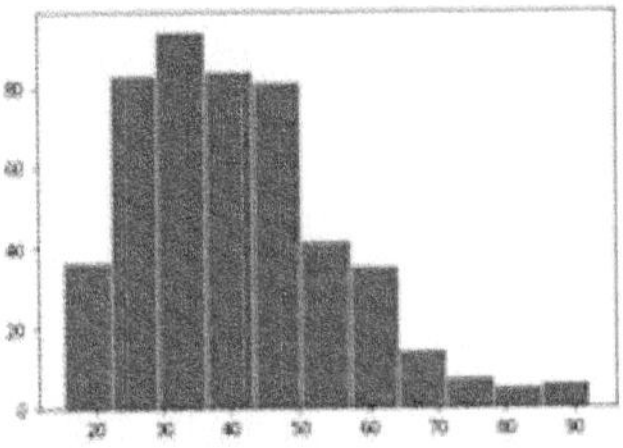	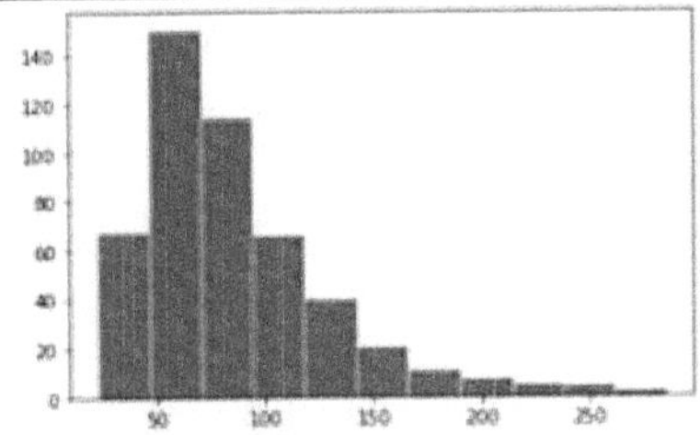

Table 5.6. *Data histograms of original data and noisy data of factor 0.11*

noise factor	**Original**	**0.11**
Gaussian distribution	X_mean = 36.898195214455285 +- 0.8603554314751484 sigma = 15.181692857262172 +- 0.9604436218495124 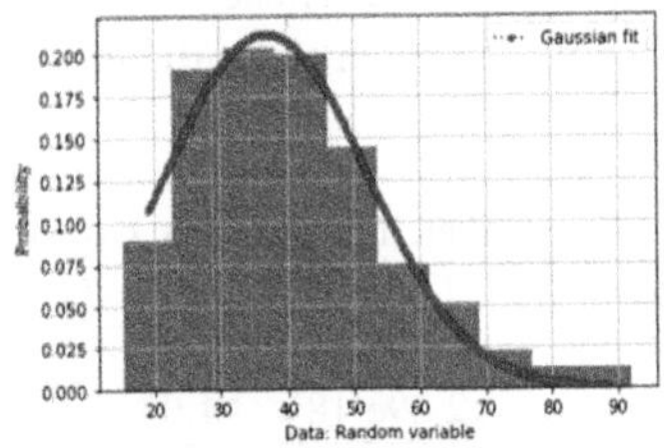	X_mean = 69.16179481189005 +- 3.337233007176847 sigma = 35.40304600717122 +- 3.792688230061604 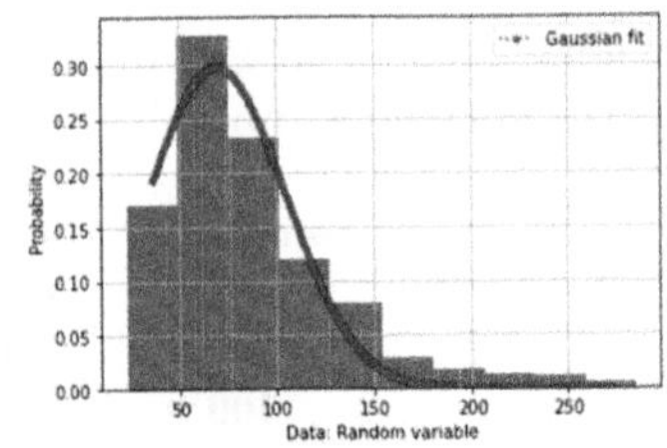

Table 5.7. *Gaussian distribution fitted to original data and noisy data of factor 0.11*

In Table 5.7, both of these datasets follow a normal probability distribution represented by their respective parameters. This information enables the switch from descriptive to inferential statistics to draw conclusions.

The Central Limit Theorem (CLT) is a fundamental concept in probability theory. It states that even though the data distribution does not fully respect the normal distribution, taking sufficiently large samples from a population will tend to result in normally distributed sample means. This holds true even though the original variables themselves are not normally distributed. In other words, the sampling distribution of the standardized sample mean tends towards the standard normal distribution for independent and identically distributed random variables.

To put it simply, the CLT proves that by increasing the sample size, the distribution of the samples becomes more and more like a bell curve. This theorem has been empirically demonstrated on our dataset, as shown in Table 5.8.

Sample size	**25**	**50**	**100**
Data histograms	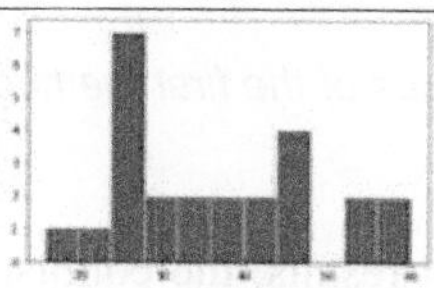	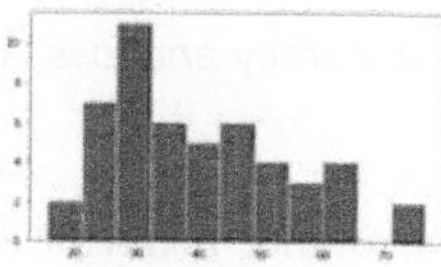	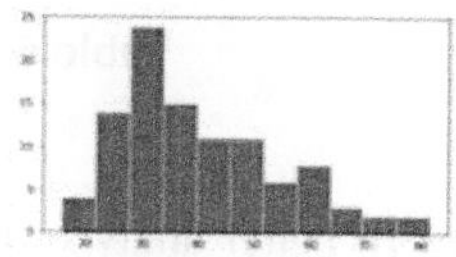

Table 5.8. *Data histograms of original data with increasing sample size*

B) **Machine learning**

To conduct experiments, the original dataset without noise will be used for the training phase of the model. A noisy dataset by salt and pepper with a factor of 0.11 will be used in batches during the production phase. The same neural network with the same configuration will be used as in the first system proposal.

5.3.2.1.3. Experiments

A) **Training phase**

Table 5.9 demonstrates the learning and loss curves, which are indicative of the model's learning progress. The learning curves depict the accuracy of the model's predictions, with the validation and training curves becoming increasingly closer as the model's accuracy improves. The loss curves, on the contrary, measure the error rate of the model. As the model's learning progresses, the validation and training loss curves become increasingly closer and lower, indicating a reduction in the model's error rate.

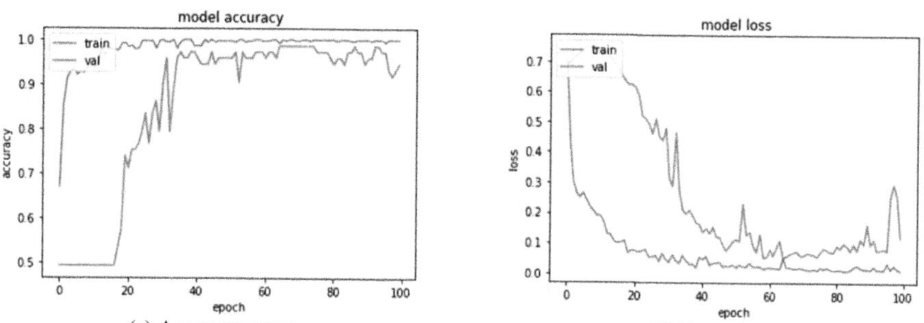

(a) Accuracy curve (b) Loss curve

Figure 5.9. *Accuracy and loss curves of the second method*

To support the results, the classifier was assessed on the testing set, and it provides satisfying results in Table 5.9.

Test accuracy	Test loss
0.9795	0.0468

Table 5.9. *Accuracy and loss values of the first method*

To better analyze the classifier behavior and results, the confusion matrix gives additional information between true and prediction labels. In Figure 5.10, negative data are correctly classified to 95% and positive data are correctly classified to 100%.

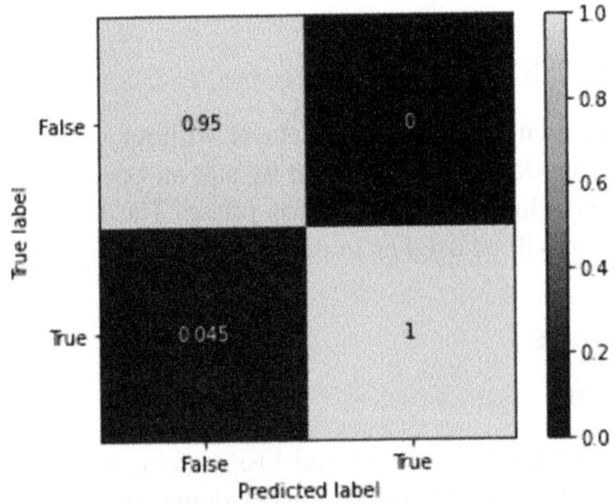

Figure 5.10. *Confusion matrix of the second method*

B) **Production phase**

(a) *Time-evolving control*

Time-evolving control is used in the machine learning (ML)-based system to control the learning process. This will result to a time-evolving learning of models on clear and noisy data.

From the previously learned model, we proceed to a learning on 50 batch.

In Figure 5.11, continuous normal and continuous noisy data is injected to show the model accuracy through time. Models that suffer from biased data show oscillating accuracy, ranging from 100% to 94%.

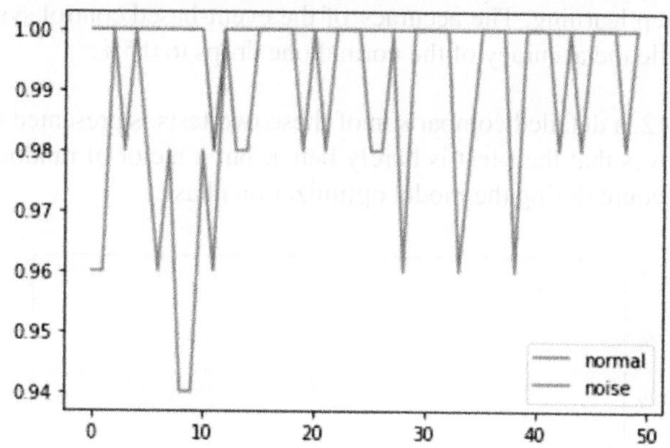

Figure 5.11. *Learning model accuracy on normal and noisy data over time*

(b) *Event-evolving control*

Event-evolving control is used in the ML-based system to control the learning process. This will result in an event-evolving learning of models on clear and noisy data. To suit real-world scenarios, instead of using continuous noisy data, episodic noisy data is injected as abnormal situations appear in real-world scenarios. Episodic data injection was done by a binary random function. From the previously learned model, we proceed to a learning on eight batches. At each step, learned or not, the model is assessed on the testing set to see the prediction degradation. In Table 5.10, the time-evolving learning system learns from episodic biased data at each step, and the event-evolving learning system learns from episodic biased data when an event is triggered.

Event trigger	**Welch t-test**	**MW U-test**
Accuracy	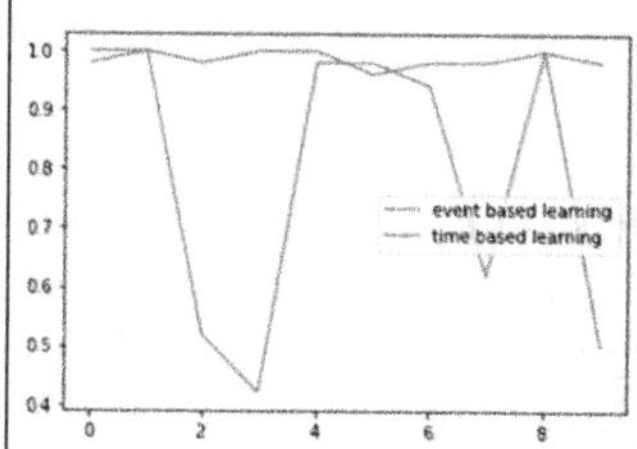	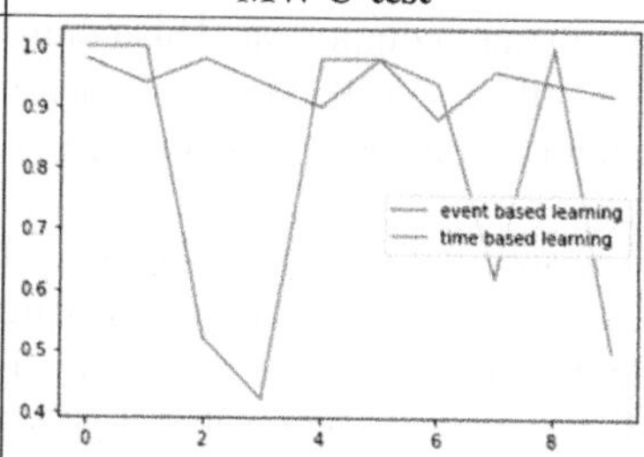

Table 5.10. *Event-based learning control with two statistical tests on episodic biased data on eight time steps*

In an evolving learning system, both event-based methods provide satisfactory results. The experiments were conducted on a longer sequence, ranging from 8 to

20 batches of step learning. The accuracy of the event-based control system remains satisfactory, while the accuracy of the normal one drops to 0.5%.

In Figure 5.12, a detailed comparison of these two tests is presented over a longer sequence. It proves that the t-test is barely better, but a factor of randomness should be taken into account during the model optimization phase.

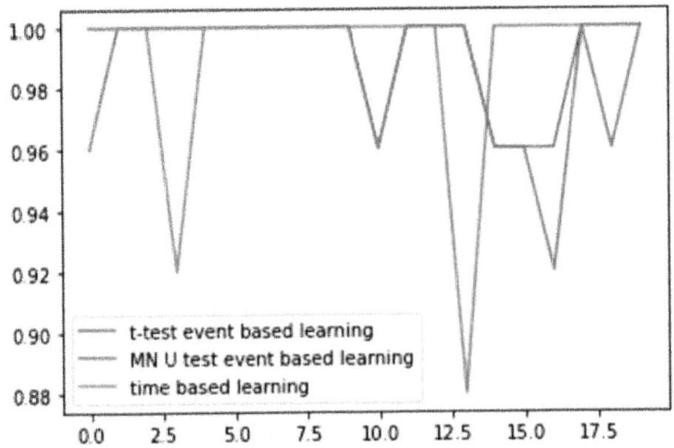

Figure 5.12. *Event-based learning control with two statistical tests on episodic biased data over 20 time steps*

5.3.2.2. *User-centric self-awareness and self-learning DSS*

5.3.2.2.1. Proposed system

In Figure 5.13, the proposed system is a control system which is triggered by an event. As in the previous methods, the control system will occupy the place of the DSS software component. This control system is composed of a prediction function which is the controlled system, a learning mechanism as a controller and a feedback generator as a measurement element.

The specificity of this approach is first in the feedback generator, which is able to measure the user satisfaction of two preferences – analyzed content and E2E perceived content – and second, in the event-triggering mechanism which is embedded in the prediction function and makes it possible to trigger an event when an abnormal VQA is detected, i.e. when video data quality is deteriorating.

This "User-centric self-awareness and self-learning DSS" approach aims to self-manage the system by considering endogenous, exogenous, and human factors that could interfere with the IVS performance. It thus gives the predictive function itself the ability to manage effectively.

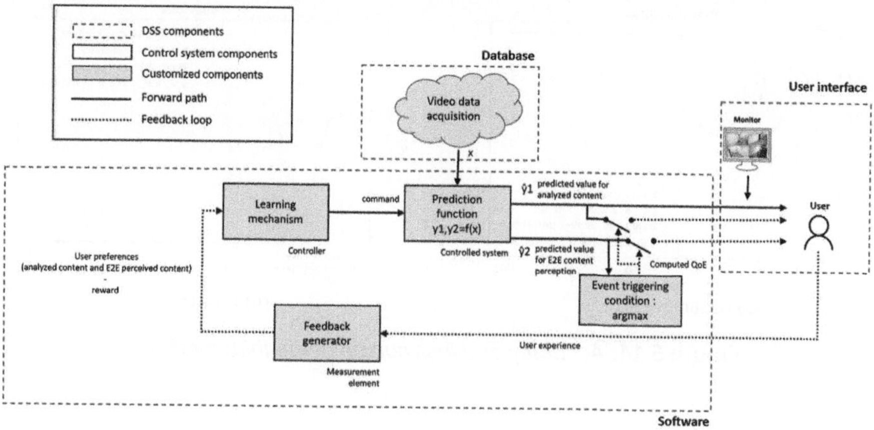

Figure 5.13. *User-centric self-aware and self-adaptive DSS*

5.3.2.2.2. Proposed method

The classifier is developed to do multitask learning for two tasks exactly, one for detecting anomaly on analyzed content and the other on E2E perceived content. A two-headed neural network or a two-branch neural network is proposed.

To conduct experiments, the dataset will be augmented by a salt and pepper noisy dataset with a factor of 0.11. In addition to the original dataset, the noisy one is created. The original dataset of 486 images distinguishes between fire and no fire and another dataset of 486 images with noise and no noise.

For the training phase, the two datasets were split into training, validation and testing set to observe the learning capacity through its learning curve and assessment on unseen data. The training ratio is set to 0.75, the validation ratio is set to 0.15 and the test ratio is set to 0.10. The learning rate is set to 0.001, and the number of epochs is set to 100.

The optimizer used is stochastic gradient descent with a momentum set to 0.9 and the decay to the initial learning rate divided by the number of epochs, so 0.001/100.

The loss function used is a binary cross-entropy which is suitable for the anomaly detection task. The classifier neural network architecture is configured with dense layers and dropout layers for regularization assignment.

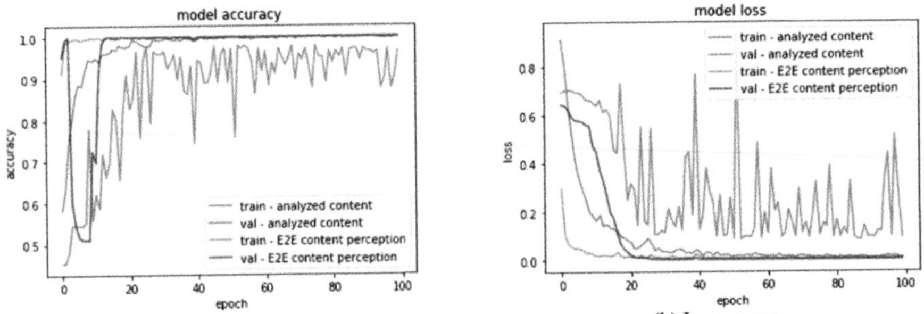

(a) Accuracy curve (b) Loss curve

Figure 5.14. *Accuracy and loss curves of the third method*

5.3.2.2.3. Experiments

A) **Training phase**

Figure 5.14 displays the learning and loss curves for both anomaly detection on analyzed content and E2E content perception. The curves show clear evidence of learning, as they progressively reach higher accuracy values with validation and training curves becoming increasingly closer. Similarly, the loss curves also provide proof of learning, as they reach lower values with validation and training loss curves that gradually become closer.

To better analyze the classifier behavior and results, the confusion matrix for analyzed content and E2E perceived content gives additional information between true and prediction labels. In Figure 5.15, for E2E perceived classifier, the model correctly classifies negative and positive labels, while for analyzed content, negative data are correctly classified to 88% and positive data are correctly classified to 100%.

Once learning proof is showed and the classifier assessment is done, our study is completed by assessing by showing real data and its output on some samples as in Table 5.11.

B) **Production phase**

In real conditions, another parameter must be taken into account; this is the learning rate because when it comes to continuously learning a model in production based on pre-knowledge acquired during the phase of training, this process is known as fine-tuning. In general, the learning rate (lr) in the production phase is lower than the lr in the training phase.

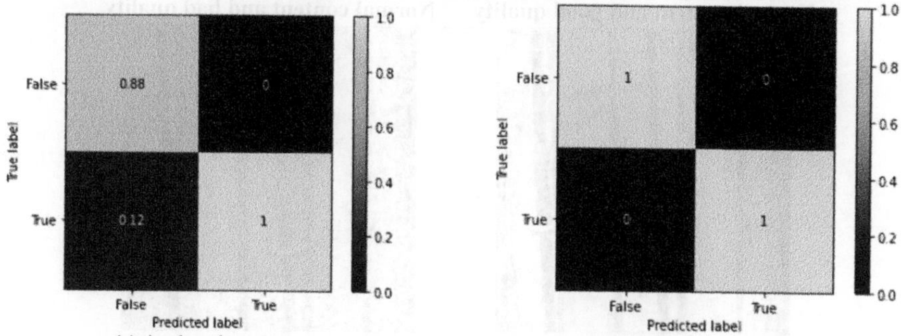

(a) Analyzed content (b) E2E perceived content

Figure 5.15. *Confusion matrix: first, fire detection and second, video quality detection*

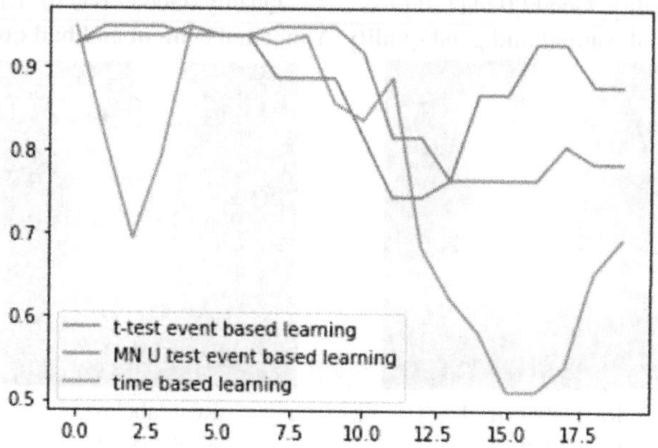

Figure 5.16. *Comparative study of event-based and time-based methods on noisy data injected randomly*

When assessing event-based learning control with statistical test-based event triggering, randomness is introduced in the optimization function at lr of 0.001, as shown in Figure 5.16. The trend in these curves is that event-based learning has better results than that which learns from all the data, called time-evolving learning.

In Figure 5.17, the results from these two experiments show that randomness during model optimization makes interpretation difficult. However, we can conclude that time-based learning consistently performs worse than all event-based learning controls. Additionally, the machine learning methods and statistical tests yield similar outcomes across the various experiments.

Normal content and good quality	**Normal content and bad quality**
	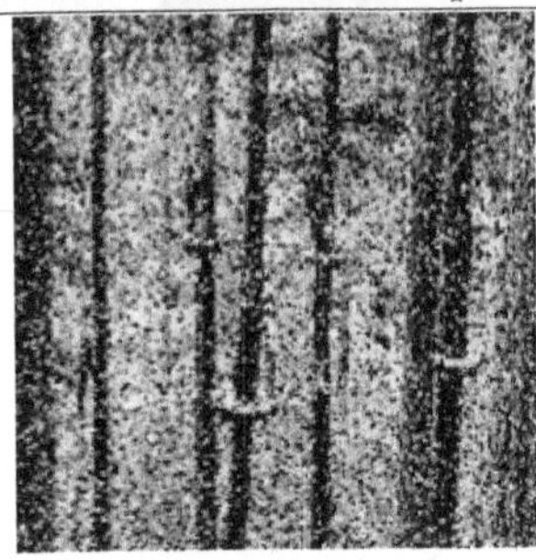
Real label o1: 0 Real label o2: 0 Pred analyzed content 0 Pred E2E content perception 0 Content: Fire/ No fire [1. 0.] Quality: Good / Bad [1. 0.]	Real label o1: 0 Real label o2: 1 Pred analyzed content 0 Pred E2E content perception 1 Content: Fire/ No fire [1. 0.] Quality: Good / Bad [0. 1.]
Abnormal content and good quality	**Abnormal content and bad quality**

Real label o1: 1 Real label o2: 0 Pred analyzed content 1 Pred E2E content perception 0 Content: Fire/ No fire [0. 1.] Quality: Good / Bad [1. 0.]	Real label o1: 1 Real label o2: 1 Pred analyzed content 1 Pred E2E content perception 1 Content: Fire/ No fire [0. 1.] Quality: Good / Bad [0. 1.]

Table 5.11. *Model output on real samples of the third method*

In Table 5.12, at lr at 0.001 and at 0.0001, in these two different experiments, the query number diverges a lot from the first to the second. The dilemma between accuracy and number of requests could be better defined here. When the number of queries increases a lot, it hardly falls into a time-scalable configuration and otherwise stays event-based, but the accuracy is not enough.

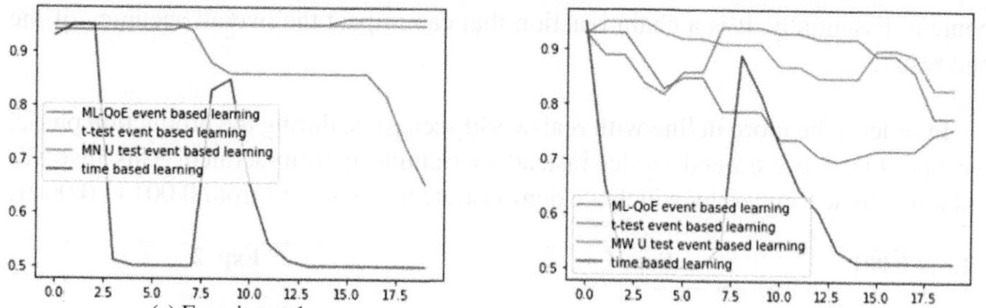

(a) Experiment 1 (b) Experiment 2

Figure 5.17. *Two experiments under identical conditions of adaptive models*

Lr	**0.001**	**0.0001**
Curves	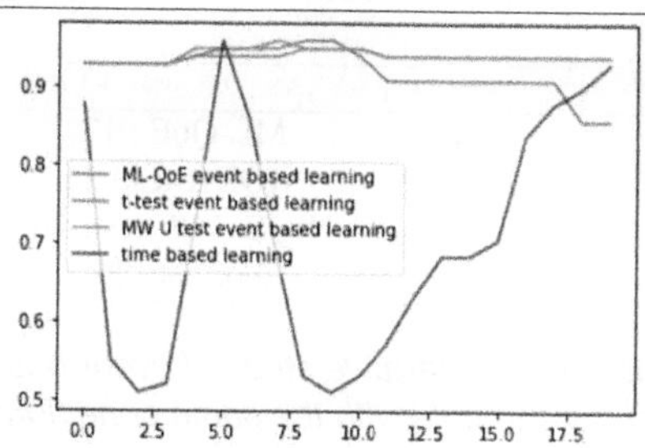	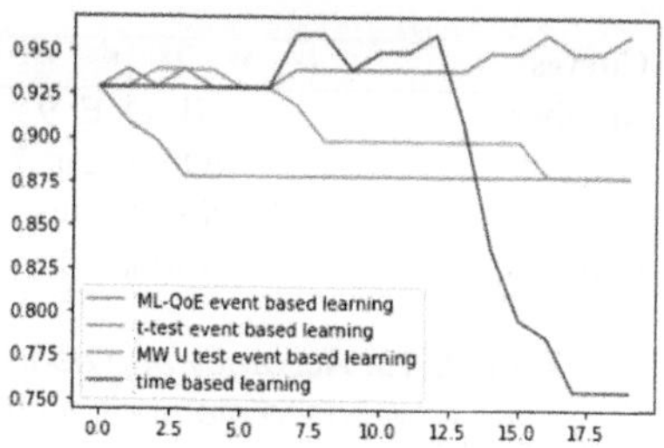
Number of QoPs user requests	ML-QoE: 7 NIQE+t-test: 5 NIQE+MW test: 5 Normal: 20	ML-QoE: 12 NIQE+t-test: 11 NIQE+MW test: 11 Normal: 20

Table 5.12. *Accuracy and user request comparative study of event-based and time-based methods on two experiments with different learning rates*

For the ML approach, queries vary a lot, but accuracy remains stable compared to statistical testing when the number of queries remains low and accuracy varies slightly. A dilemma exists between accuracy and the number of requests to reduce. This opens the discussion between the needs of the end-user.

Unfortunately, as previously stated and shown in Table 5.13, at the same lr at 0.001, two different experiments were performed and the event trigger leads to differences in accuracy and requests of the user.

The conclusion in Table 5.13 suggests that the accuracy of the analyzed content is affected when the perceived E2E content accuracy is not achieved. This happens because the triggering event method relies on the correctness of the content perceived by E2E, and any bias in that data can result in biased learning and inaccurate analyzed content. Essentially, it is a chain reaction that can impact the overall accuracy of the end result.

In order to be more in line with real-world scenarios, during the production phase, we fine-tune a pre-trained model instead of training it from scratch. This involves reducing the learning rate, which we demonstrate by switching from 0.001 to 0.0001.

Lr at 0.001	Exp. 1	Exp. 2
Curves	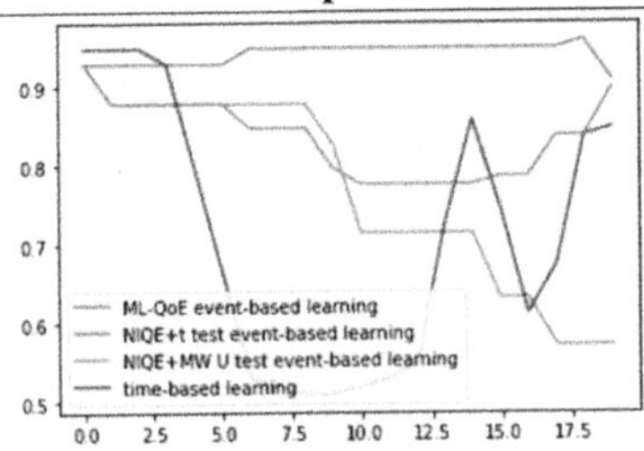	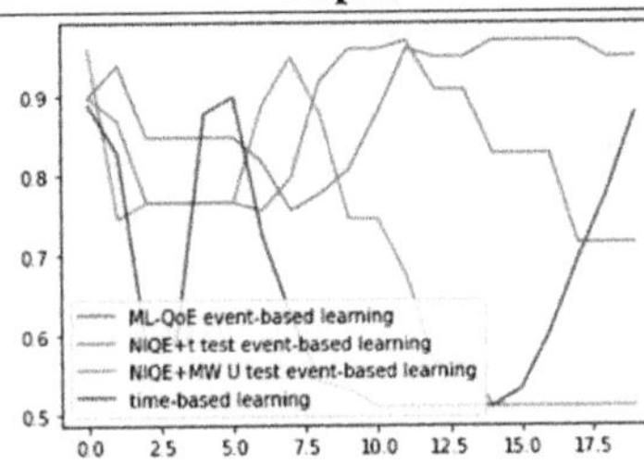
Number of QoPs user requests	ML-QoE: 9 NIQE+t-test: 7 NIQE+MW test: 7 Normal: 20	ML-QoE : 17 NIQE+t-test: 11 NIQE+MW test: 10 Normal: 20

Table 5.13. *Accuracy and user request comparative study of event-based and time-based methods on two experiments with the same conditions*

Accuracy curve	Number of QoPs user request
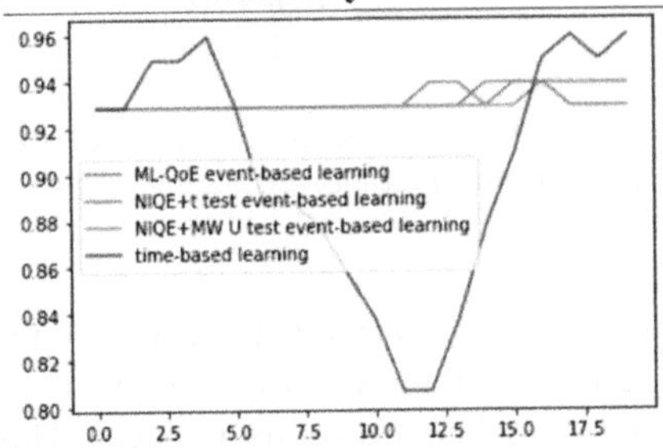	ML-QoE: 8 NIQE+t-test: 7 NIQE+MW test: 7 Normal: 20

Table 5.14. *Accuracy and user request comparative study of event-based and time-based methods at lr = 0.0001 and noise = 0.11*

In Table 5.14, the results may be questionable because the model was trained to generalize over the noise at 0.11, so it could detect fire or not inside this noise. A noise of 0.4 as in Figure 5.18 is provided inside the system to assess the system output and reaction to it.

Figure 5.18. *Sample of noise level of 0.4 applied to the system*

To assess the impact of any changes, increasing the learning rate is an effective way to observe it. In order to do this, we set the learning rate back to its original value of 0.001. The experiment was performed at lr = 0.001 with unseen noisy data at 0.4, as shown in Table 5.15. This unseen noisy data was not used during the training phase, but it will be introduced during the production phase to observe how the adaptive model reacts to it.

Accuracy curve	Number of QoPs user request
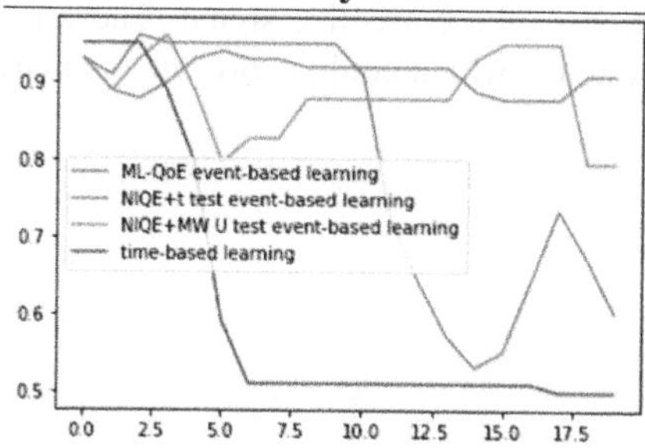	ML-QoE: 18 NIQE+t-test: 10 NIQE+MW test: 10 Normal: 20

Table 5.15. *Accuracy and user request comparative study of event-based and time-based methods at lr = 0.001 and noise = 0.4*

In Table 5.15, it shows that the accuracy of the model decreases and the accuracy of our Quality of Experience (QoE)-based event detection also decreases and the Quality of Prediction (QoP) user queries for the ML method increases up to 18 times. Indeed, we have not learned to detect this noise level as abnormal where statistical models based on the Natural Image Quality Evaluator (NIQE) perform better.

The accuracy of our model drops significantly when the QoE predictor fails and reaches almost the same amount of QoP queries as standard time-based learning would.

5.4. Conclusion

In this chapter, we present a new autonomic approach for an IVS system. The purpose of this approach is to enable the IVS system to self-manage, thereby reducing the time required for user management. This, in turn, will increase user acceptability. To develop this approach, we drew inspiration from control theory and identified two types of approach – time-based and event-based.

Three DSS are proposed based on ML, descriptive statistics and inferential statistics. The first system does not require an event trigger and explains how to deploy an autonomic IVS for self-configuration. The second system triggers an event when the video quality is not suitable. This is done using descriptive and inferential statistics, which draw some conclusions about the samples and allow the system to self-protect from noise injection. The third system triggers an event when the video quality is not suitable to learn, and we will enable the ML algorithm to be aware of the situation on its own to trigger its learning.

These systems demonstrate the application of an autonomic system in video surveillance and the potential application of the corrective signal. These proposals are part of the premise of autonomic systems, which could evolve in an environment with reduced user supervision.

Conclusions and Perspectives

"Small aim is a crime; have great aim."
A.P.J. Abdul Kalam

C.1. Summary of contributions

Smart cities will benefit from next-generation sensors, communication and the Internet, which will make decision support systems (DSS) more efficient. The new generation systems will provide more data through video surveillance, but it is important that they are relevant to decision-makers.

First, we proposed a literature review on a specific topic, "An End-to-End Video Surveillance System for Public Safety". This will help us analyze human threats to homeland security in both the physical and cyber spaces.

Second, we proposed a framework that addresses the challenges of handling vast amounts of information from multiple sources, working in a high-pressure environment, dealing with complex tasks and making critical decisions. Unlike existing methods, our contribution allows for the deployment of a DSS in real-world scenarios using reinforcement signals and quality of experience (QoE). This will keep the situational awareness (SA) continuously up to date by leveraging user expertise. Our approach introduces a context-evolving system for mission-critical operations that takes into account the human factor. In this system, a closed loop will help control errors.

Third, we suggested a corrective signal for the system that considers various criteria, including endogenous, exogenous and human factors. For this, we use a task-specific QoE metric that reflects the user's perspective. To calculate this metric, we rely on a context enriched with different metrics, such as QoP for video analytics-based situation assessment, traditional QoE for video quality evaluation and augmented QoE for assessing issues with camera obstruction and video integrity. This study indicates research directions for further improvement.

Fourth, our research is dedicated to developing capabilities and result-oriented solutions for abnormal situation detection and creating applications that meet mission-critical requirements for real-world scenarios. To achieve this, we proposed predictive systems that address various issues, including model interpretability, temporal anomaly localization, trial-and-error learning and context drift. We developed four proposals, of which the first two focus on improving the capabilities of the model, while the last two aim at improving performance. The first proposal combines a clustering model with statistical models SOM and HMM to enable temporal anomaly localization and the interpretability of multi-instance learning, which drives the decision-making process. The second proposal involves a neural network that uses a reinforcement learning algorithm called actor-critic to address temporal anomaly localization, concept drift and trial-and-error learning problems by continuously learning with rewards. Our third approach combines the previous two methods, using SOM for clustering and an FFNN for the policy-gradient RL algorithm. This proposal addresses temporal anomaly localization and trial-and-error learning but deals with concept drift problems and interpretability less effectively. It provides better results than previous research but still needs improvement in outcomes. Our fourth proposal introduces a novel framework for multi-instance anomaly detection, where a new learning scheme for neural networks is proposed. This method, called "similarity-based meta-reinforcement learning", allows for dealing with temporal anomaly localization, concept drift, trial-and-error learning and interpretability. This approach can also deal with fast context change to learn quickly and provide a result of 79.9% AUC. This last approach enables our SA system to perform online adaptive learning, develop a long-term vision and strategy, and efficiently localize anomalies temporally with competitive results. To simplify the process and reduce user requests, our system employs reward-based multi-instance learning to minimize labeling and label complexity. In order to comply with GDPR regulations, our end-to-end framework refrains from using individual analysis and instead learns through trial and error from new experiences while restricting data access to build prior knowledge. Finally, the learning and inference algorithm satisfies real-time requirements.

Fifth, we also conducted research on the overall system management to reduce product ownership and increase performance, where the user will be a constraint, for example, in terms of reaction time and remediation, configuration time and human resource allocation. Therefore, we designed a self-managed intelligent video surveillance system with end-to-end trustworthiness from content perception to content analysis. To achieve this, we designed and experimented with a user-centric self-configuring DSS to save the user configuration time at each installation, a user-centric self-protecting DSS to protect the system from intentional or unintentional data source deterioration and a user-centric self-awareness and self-learning DSS to self-manage the system from endogenous, exogenous and human factors that could interfere with its performance. These three systems provide satisfying preliminary results, demonstrate their application potential and could be further developed in the future.

C.2. Perspectives and future work

Some improvements are required in terms of RL algorithm convergence, speed of learning and data gathering. Further assessment is necessary for the anticipation component, and the anomaly detection needs to be developed to reach a better level of convergence. The current level of convergence was sufficient for a proof of concept, but there is potential for achieving better results with a few experiments. With a new dataset, anomaly detection on the analyzed content and E2E perceived content should provide further insight into anomaly detection in cyber and physical space. Additionally, more work is needed on the interpretability part and trial-and-error learning, which can considerably increase the training time and lead to issues such as gradient exploding and other drawbacks.

According to promising results, this work can be extended with various research perspectives as follows.

– Time-evolving system in real-world conditions

- Feedback requirements: feedback plays a crucial role in the development of AI algorithms. It enables them to learn and improve their performance over time. By receiving feedback on their decisions and actions, AI algorithms can modify their approach and enhance their accuracy and efficiency. Feedback can be in the form of labeled data, user input or performance metrics. For example, continuous reinforcement learning (RL) has shown great efficiency in virtual simulated environments, where it is possible to start learning with dense rewards and then switch to sparse rewards. However, in the real world, the goal is to move away from greedy algorithms and explore different approaches, such as using heuristics or dealing with the credit assignment problem by employing MIL or RL techniques. There is a need to explore the possibility of learning from a large quantity of data with only a few metadata.

- Learning time: real-time decision-making systems require accurate and prompt decisions. If an AI model takes too long to learn and make decisions, it may not be able to keep up with the demands of real-time decision-making. This can result in delayed or incorrect decisions that may have far-reaching consequences in various industries. Fast context adaptation affords the ability to learn quickly in novel contexts. In order to achieve fast convergence to the anticipated solution, a smooth transition approach may be investigated, which could involve the manipulation of hyperparameters, such as the learning rate, or the enhancement of generalizing capabilities via its internal components, using similarity, meta, reinforcement or multi-instance learning, for instance.

- Unnecessary exploration: exploration plays a critical role in reinforcement learning as it helps to comprehend the environment better. If exploration is not carried out, the algorithm might become too focused on short-term rewards and overlook potential long-term strategies. However, excessive exploration can also be problematic because it can cause the agent to spend too much time exploring insignificant or irrelevant parts of the environment, leading to slower learning and suboptimal performance. Hence, it is essential to strike a balance between exploration and exploitation to ensure efficient and effective learning. Moreover, it is crucial to reduce computer power needs and ensure energy efficiency for long-term performance and power consumption.

- Unnecessary rewarding: it is vital to understand that AI systems can experience an overload of data due to unnecessary rewards. This can result in increased processing times and resource consumption, ultimately having a detrimental effect on the learning process. In order to prevent such situations, it is imperative to provide only relevant and meaningful feedback to the AI system. Additionally, reducing user requests is a critical aspect of optimizing the system's performance. Event-driven rewarding processes are an effective approach to achieving this, reducing the need for user management time. To ensure the safety of AI, an uncertainty-driven rewarding process can be implemented, which guarantees an autonomous process until user supervision is required. Novelty detection techniques can also be employed to detect when the system deviates from its initial detection domain. However, the development of this field of study is still necessary.

– End-to-end system trustworthiness

- Information chain: in order to make well-informed decisions based on reliable information, it is imperative to ensure the trustworthiness of the information being disseminated. This can be achieved by employing cognitive biases to control the information chain. Furthermore, it is essential to adhere to the quality of service (QoS) paradigm, which involves a set of monitoring and controlling processes that ensure the technical requirements of the solution are met.

- Decision chain: however, trustworthiness of the information alone is insufficient to guarantee well-informed decision-making. The trustworthiness of the reasoning mechanism itself is critical to this process. To ensure this, it is necessary to control the decision chain and explore research topics that cater to the user's demands. This includes developing interpretations or explainable algorithms that aid in understanding the reasoning behind the decision.

- End-to-end flow: it is crucial to evaluate the entire data chain from start to finish to gain the right insights for making well-informed decisions. This helps to understand the bigger picture and overall context that impacts the decision. It is essential to capture user expectations at the end of the information chain as it leads to a better understanding of the user's preferences. Ensuring trustworthy decision-making throughout the entire process is necessary. To achieve this, it is recommended to investigate a certified process to validate the information and decision chain either simultaneously or separately.

– Advanced knowledge discovery

- Raw data: in order to ensure that users' expectations (Y in ML terminology) are in line with the context (X), it is essential to collect QoE data to understand user goals. However, this process may be hindered by a lack of context. Sufficient context is required to align user expectations with the context, which in turn may lead to the discovery of new patterns. Nevertheless, based on user expectations, this may not always be necessary. The utilization of big data and data mining concepts can enhance the effectiveness of data enrichment techniques.

- Goal: to ensure that user expectations (Y in ML or Q in MCDM terminologies) are met, it is crucial to gather context. Several studies have been conducted on improving QoE by identifying the factors that impact it and resolving issues. However, instead of relying solely on automatically collecting specific metrics to build QoE, it is recommended to use QoE discovery to model complex human beliefs, as advocated in human-centered AI.

- Human intelligence mimics: in an effort to enhance machine performance, researchers have dedicated significant attention to achieving human-like intelligence. To achieve this objective, they have employed energy-intensive algorithms, boosted computing power, and relied on large datasets. However, the pursuit of replicating human cognitive abilities may also lead to improved system capabilities. In particular, it could address contemporary challenges such as data scarcity, explainability, trustworthy systems, memory and rapid learning. This could result in systems that are better able to handle complex tasks and offer a range of benefits.

References

Abadi, M., Agarwal, A., Barham, P., Brevdo, E., Chen, Z., Citro, C., Corrado, G.S., Davis, A., Dean, J., Devin, M. et al. (2015). TensorFlow: Large-scale machine learning on heterogeneous systems. [Online]. Available at: https://www.tensorflow.org/.

Abar, T., Letaifa, A., Asmi, S.E. (2022). QoE-based intelligent intrusion detection use case: University video surveillance. In *2022 IEEE 9th International Conference on Sciences of Electronics, Technologies of Information and Telecommunications (SETIT)*, 7–11 [Online]. Available at: https://api.semanticscholar.org/CorpusID:252164738.

Aberkane, S. and Elarbi, M.S. (2019). Deep reinforcement learning for real-world anomaly detection in surveillance videos. In *2019 6th International Conference on Image and Signal Processing and their Applications (ISPA)*, 1–5 [Online]. Available at: https://api.semanticscholar.org/CorpusID:210931595.

Aboah, A., Shoman, M., Mandal, V., Davami, S., Adu-Gyamfi, Y., Sharma, A. (2021). A vision-based system for traffic anomaly detection using deep learning and decision trees. In *2021 IEEE/CVF Conference on Computer Vision and Pattern Recognition Workshops (CVPRW)*, 4202–4207.

Airbus DS SLC. (n.d.). Agnet [Online]. Available at: https://www.securelandcommunications.com/agnet [Accessed 25 September 2023].

amNY (2022). September sees continued rise in most major crimes in New York City [Online]. Available at: https://www.amny.com/news/crime-rates-new-york-city-september-2022/ [Accessed 25 June 2023].

Anala, M.R., Makker, M., Ashok, A. (2019). Anomaly detection in surveillance videos. In *2019 26th International Conference on High Performance Computing, Data and Analytics Workshop (HiPCW)*, 93–98 [Online]. Available at: https://api.semanticscholar.org/CorpusID:211244682.

Bagdasaryan, E., Veit, A., Hua, Y., Estrin, D., Shmatikov, V. (2018). How to backdoor federated learning [Online]. Available at: https://api.semanticscholar.org/CorpusID:49557410.

Baum, L.E. and Petrie, T. (1966). Statistical inference for probabilistic functions of finite state Markov chains. *The Annals of Mathematical Statistics*, 37(6), 1554–1563.

Belmondo, C. (2006). Ressources attentionnelles et processus de création de connaissances. In *Capital immatériel, connaissance et performance*, Bounfour, A. (ed.). L'Harmattan, Paris.

Bhatnagar, S., Sutton, R.S., Ghavamzadeh, M., Lee, M. (2009). Natural actor-critic algorithms. *Automatica*, 45, 2471–2482 [Online]. Available at: https://api.semanticscholar.org/CorpusID:13564930.

Bosch (n.d.). Bosch IP camera vulnerability [Online]. Available at: https://psirt.bosch.com/security-advisories/bosch-2018-1202.html [Accessed 15 May 2021].

Bowling, B. and Iyer, S. (2019). Automated policing: The case of body-worn video. *International Journal of Law in Context*, 15, 140–161 [Online]. Available at: https://api.semanticscholar.org/CorpusID:197704529.

Cámara, J., Bellman, K.L., Kephart, J., Autili, M., Bencomo, N., Diaconescu, A., Giese, H., Götz, S., Inverardi, P., Kounev, S. et al. (2017). Self-aware computing systems: Related concepts and research areas [Online]. Available at: https://api.semanticscholar.org/CorpusID:36654587.

Carreira, J. and Zisserman, A. (2017). Quo vadis, action recognition? A new model and the kinetics dataset. In *2017 IEEE Conference on Computer Vision and Pattern Recognition (CVPR)*, Honolulu, HI, 4724–4733.

CES Complete (n.d.). Deal-cctv-camera-vandalism [Online]. Available at: https://cescomplete.com/2017/12/18/deal-cctv-camera-vandalism/ [Accessed 25 June 2023].

Chaaraoui, A.A., Climent-Pérez, P., Flórez-Revuelta, F. (2012). A review on vision techniques applied to human behaviour analysis for ambient-assisted living. *Expert Systems with Applications*, 39, 10873–10888 [Online]. Available at: https://api.semanticscholar.org/CorpusID:27637541.

Chen, N., Chen, Y., Song, S., Huang, C.-T., Ye, X. (2016). Poster abstract: Smart urban surveillance using fog computing. In *2016 IEEE/ACM Symposium on Edge Computing (SEC)*, Washington, D.C., 95–96.

Chen, Y.Y., Liu, Z., Zhang, B., Fok, W.W.T., Qi, X., Wu, Y.-C. (2022). MGFN: Magnitude-contrastive glance-and-focus network for weakly-supervised video anomaly detection [Online]. Available at: https://api.semanticscholar.org/CorpusID:254044120.

CSO Online (n.d.). The Mirai botnet explained: How teen scammers and CCTV cameras almost brought down the internet [Online]. Available at: https://www.csoonline.com/article/3258748/the-mirai-botnet-explained-how-teen-scammers-and-cctv-cameras-almost-brought-down-the-internet.html [Accessed 15 May 2021].

CyberScoop (2017). D.C. police department hit with ransomware [Online]. Available at: https://www.cyberscoop.com/washington-dc-ransomware-dc-police-department/ [Accessed 25 June 2023].

Degardin, B. and Proença, H. (2020). Human activity analysis: Iterative weak/self-supervised learning frameworks for detecting abnormal events. In *2020 IEEE International Joint Conference on Biometrics (IJCB)*, 1–7 [Online]. Available at: https://api.semanticscholar.org/CorpusID:230994351.

Djeachandrane, A., Hoceini, S., Delmas, S., Duquerrois, J.-M., Mellouk, A. (2022). QoE-based situational awareness-centric decision support for network video surveillance. In *ICC 2022 – IEEE International Conference on Communications*, 335–340 [Online]. Available at: https://api.semanticscholar.org/CorpusID:251535115.

Endsley, M.R. (1995). Toward a theory of situation awareness in dynamic systems. *Human Factors: The Journal of Human Factors and Ergonomics Society*, 37, 32–64.

Endsley, M.R., Bolte, B., Jones, D.G. (2003). *Designing for Situation Awareness: An Approach to User-Centered Design*. Taylor & Francis, Abingdon-on-Thames.

Federal Aviation Administration (1993). Human factors policy [Online]. Available at: https://www.faa.gov/documentLibrary/media/Order/9550.8.pdf [Accessed 21 September 2023].

Feng, J., Zhang, C., Hao, P. (2010). Online learning with self-organizing maps for anomaly detection in crowd scenes. In *2010 20th International Conference on Pattern Recognition*, 3599–3602 [Online]. Available at: https://api.semanticscholar.org/CorpusID:14445229.

Feng, J., Hong, F.-T., Zheng, W. (2021). Mist: Multiple instance self-training framework for video anomaly detection. In *2021 IEEE/CVF Conference on Computer Vision and Pattern Recognition (CVPR)*, 14004–14013 [Online]. Available at: https://api.semanticscholar.org/CorpusID:233025083.

French Ministry (2023). Insécurité et délinquance en 2022 : une première photographie – interstats analyse n°54 [Online]. Available at: https://www.interieur.gouv.fr/Interstats/Actualites/Insecurite-et-delinquance-en-2022-une-premiere-photographie-Interstats-Analyse-N-54 [Accessed 25 June 2023].

Galteri, L., Seidenari, L., Uricchio, T., Bertini, M., Bimbo, A. (2020). Preserving low-quality video through deep learning. *IOP Conference Series: Materials Science and Engineering*, [Online]. Available at: https://api.semanticscholar.org/CorpusID:228844620.

Gama, J., Žliobaitė, I., Bifet, A., Pechenizkiy, M., Bouchachia, A. (2014). A survey on concept drift adaptation. *ACM Computing Surveys (CSUR)*, 46, 1–37 [Online]. Available at: https://api.semanticscholar.org/CorpusID:201087802.

Gamaleldin, A., Atef, A., Saker, H., Shaheen, A. (n.d.). Fire dataset [Online]. Available at: https://www.kaggle.com/datasets/phylake1337/fire-dataset.

Ganek, A.G. and Corbi, T.A. (2003). The dawning of the autonomic computing era. *IBM Systems Journal*, 42, 5–18 [Online]. Available at: https://api.semanticscholar.org/CorpusID:8455119.

Gao, Y., Cao, H., Xu, S., Zhang, J., Ning, X., Ma, X., Adu, B.S. (2018). Anomaly detection for videos of crowded scenes based on optical flow information. In *2018 3rd International Conference on Advanced Robotics and Mechatronics (ICARM)*, Singapore, 869–879.

Gauss, C.F. (2014). Theoria motus corporum coelestium in sectionibus conicis solem ambientium [Online]. Available at: https://api.semanticscholar.org/CorpusID:61133059.

Geraldes, R., Gonçalves, A., Lai, T., Villerabel, M., Deng, W., Salta, A., Nakayama, K., Matsuo, Y., Prendinger, H. (2019). UAV-based situational awareness system using deep learning. *IEEE Access*, 7, 122583–122594.

Ghadiyaram, D., Chen, C., Inguva, S., Kokaram, A.C. (2017). A no-reference video quality predictor for compression and scaling artifacts. In *2017 IEEE International Conference on Image Processing (ICIP)*, 3445–3449 [Online]. Available at: https://api.semanticscholar.org/CorpusID:3463873

Götzinger, M., Juhász, D., Taherinejad, N., Willegger, E., Tutzer, B., Liljeberg, P., Jantsch, A., Rahmani, A.-M. (2020). Rosa: A framework for modeling self-awareness in cyber-physical systems. *IEEE Access*, 8, 141373–141394 [Online]. Available at: https://api.semanticscholar.org/CorpusID:221120586.

Grahl, M., Spring, A., Andreeva, T., Bluhm, T., Bozhenkov, S., Dumke, S., Geiger, J., Grulke, O., Grün, M., Holtz, A. et al. (2020). W7-X logbook REST API for processing experimental metadata and data enrichment at the Wendelstein 7-X stellarator. *Fusion Engineering and Design*, 160, 111819 [Online]. Available at: https://api.semanticscholar.org/CorpusID:224964713.

Grundstein, M. and Rosenthal-Sabroux, C. (2001). Vers un système d'information source de connaissances. *Ingénierie des systèmes d'information, Hermes*, 317–348.

Hao, Y., Xu, Z., Liu, Y., Wang, J., Fan, J.-L. (2018). Effective crowd anomaly detection through spatio-temporal texture analysis. *International Journal of Automation and Computing*, 16, 27–39.

Hasan, M., Choi, J., Neumann, J., Roy-Chowdhury, A.K., Davis, L.S. (2016). Learning temporal regularity in video sequences. In *2016 IEEE Conference on Computer Vision and Pattern Recognition (CVPR)*, 733–742 [Online]. Available at: https://api.semanticscholar.org/CorpusID:2429016.

Höferlin, B., Höferlin, M., Michael, R., Heidemann, G., Weiskopf, D. (2011). Interactive auditory display to support situational awareness in video surveillance. *Proceedings of the International Conference on Auditory Display* [Online]. Available at: https://api.semanticscholar.org/CorpusID:887317

Horn, P.J. (2001). Autonomic computing: IBM's perspective on the state of information technology [Online]. Available at: https://api.semanticscholar.org/CorpusID:59693806.

Huebscher, M.C. and Mccann, J.A. (2008). A survey of autonomic computing – Degrees, models, and applications. *ACM Computing Surveys*, 40, 7:1–7:28. Available at: https://api.semanticscholar.org/CorpusID:8914588.

Human Factors and Ergonomics Society. (n.d.). What is human factors and ergonomics? [Online]. Available at: https://www.hfes.org/About-HFES/What-is-Human-Factors-and-Ergonomics [Accessed 21 September 2023].

Itseez (2000). Opencv [Online]. Available at: https://github.com/opencv/opencv.

Jeftenić, N., Simić, M., Stamenkovic, Z. (2020). Impact of environmental parameters on SNR and RSS in LoRaWAN. In *2020 International Conference on Electrical, Communication, and Computer Engineering (ICECCE)*, 1–6 [Online]. Available at: https://api.semanticscholar.org/CorpusID:221387834.

Jones, D.G. and Endsley, M.R. (1996). Sources of situation awareness errors in aviation. *Aviation, Space, and Environmental Medicine*, 67(6), 507–512.

Juneja, M. and Sandhu, P. (2009). Designing of robust image steganography technique based on LSB insertion and encryption. In *2009 International Conference on Advances in Recent Technologies in Communication and Computing*, Kerala, 302–305.

KaewTrakulPong, P. and Bowden, R. (2002). An improved adaptive background mixture model for real-time tracking with shadow detection. In *Video-Based Surveillance Systems: Computer Vision and Distributed Processing*, Remagnino, P., Jones, G.A., Paragios, N., Regazzoni, C.S. (eds). Springer, Boston, MA.

Kazemi, V. and Sullivan, J. (2014). One millisecond face alignment with an ensemble of regression trees. In *2014 IEEE Conference on Computer Vision and Pattern Recognition*, 1867–1874 [Online]. Available at: https://api.semanticscholar.org/CorpusID:2031947.

Kephart, J. (2005). Research challenges of autonomic computing. In *27th International Conference on Software Engineering, 2005. ICSE 2005*, 15–22 [Online]. Available at: https://api.semanticscholar.org/CorpusID:11103966.

Kim, B. and Lee, J. (2019). A video-based fire detection using deep learning models. *Applied Sciences* [Online]. Available at: https://api.semanticscholar.org/CorpusID:199655083.

King, D.E. (2009). Dlib-ml: A machine learning toolkit. *Journal of Machine Learning Research*, 10, 1755–1758 [Online]. Available at: https://api.semanticscholar.org/CorpusID:6155330.

Kohonen, T. (2004). Self-organized formation of topologically correct feature maps. *Biological Cybernetics*, 43, 59–69 [Online]. Available at: https://api.semanticscholar.org/CorpusID:206775459.

Korshunov, P. and Ooi, W.T. (2011). Video quality for face detection, recognition, and tracking. *ACM Transactions on Multimedia Computing, Communications, and Applications*, 7, 14:1–14:21 [Online]. Available at: https://api.semanticscholar.org/CorpusID:16439190

Lamé, A. (2019). Gaze tracking [Online]. Available at: https://github.com/antoinelame/GazeTracking.

Laugel, T. (2020). Interprétabilité locale post-hoc des modèles de classification boites noires. PhD Thesis, Sorbonne Université, Paris [Online]. Available at: http://www.theses.fr/2020SORUS215.

Li, Y. (2017). Deep reinforcement learning: An overview [Online]. Available at: https://api.semanticscholar.org/CorpusID:17540505.

Li, T., Chang, H., Wang, M., Ni, B., Hong, R., Yan, S. (2015). Crowded scene analysis: A survey. *IEEE Transactions on Circuits and Systems for Video Technology*, 25(3), 367–386.

Li, T., Wang, Z., Liu, S., Lin, W.-Y. (2021). Deep unsupervised anomaly detection. In *2021 IEEE Winter Conference on Applications of Computer Vision (WACV)*, 3635–3644 [Online]. Available at: https://api.semanticscholar.org/CorpusID:231860720.

Liao, P.-S., Chen, T.-S., Chung, P.C. (2001). A fast algorithm for multilevel thresholding. *Journal of Information Science and Engineering*, 17, 713–727 [Online]. Available at: https://api.semanticscholar.org/CorpusID:9609430.

Lin, G., Zhang, Y., Xu, G., Zhang, Q. (2019). Smoke detection on video sequences using 3D convolutional neural networks. In *Fire Technology*, 1–21 [Online]. Available at: https://api.semanticscholar.org/CorpusID:117172469.

Liu, Z., Lin, Y., Cao, Y., Hu, H., Wei, Y., Zhang, Z., Lin, S., Guo, B. (2021). Swin transformer: Hierarchical vision transformer using shifted windows. In *2021 IEEE/CVF International Conference on Computer Vision (ICCV)*, Montreal, 9992–10002.

López-López, E., Regueiro, C.V., Pardo, X.M., Franco, A., Lumini, A. (2019). Incremental learning techniques within a self-updating approach for face verification in video-surveillance. In *Iberian Conference on Pattern Recognition and Image Analysis* [Online]. Available at: https://api.semanticscholar.org/CorpusID:202728737.

Lu, C., Shi, J., Jia, J. (2013). Abnormal event detection at 150 FPS in MATLAB. In *2013 IEEE International Conference on Computer Vision*, 2720–2727 [Online]. Available at: https://api.semanticscholar.org/CorpusID:6070091.

Lv, H., Zhou, C., Cui, Z., Xu, C., Li, Y., Yang, J. (2020). Localizing anomalies from weakly-labeled videos. *IEEE Transactions on Image Processing*, 30, 4505–4515 [Online]. Available at: https://api.semanticscholar.org/CorpusID:221186800.

van der Maaten, L. and Hinton, G.E. (2008). Visualizing data using t-SNE. *Journal of Machine Learning Research*, 9, 2579–2605 [Online]. Available at: https://api.semanticscholar.org/CorpusID:5855042.

Mahadevan, V., Li, W., Bhalodia, V., Vasconcelos, N. (2010). Anomaly detection in crowded scenes. In *2010 IEEE Computer Society Conference on Computer Vision and Pattern Recognition*, San Francisco, 1975–1981.

Marfievici, R., Murphy, A.L., Picco, G.P., Ossi, F., Cagnacci, F. (2013). How environmental factors impact outdoor wireless sensor networks: A case study. In *2013 IEEE 10th International Conference on Mobile Ad-Hoc and Sensor Systems*, 565–573 [Online]. Available at: https://api.semanticscholar.org/CorpusID:18677085.

Mehran, R., Oyama, A., Shah, M. (2009). Abnormal crowd behavior detection using social force model. In *2009 IEEE Conference on Computer Vision and Pattern Recognition*, Miami, 935–942.

Mellouk, A., Hoceini, S., Tran, H. (2013). *Quality of Experience for Multimedia: Application to Content Delivery Network Architecture*. ISTE Ltd, London and John Wiley & Sons, New York.

Militello, L.G., Arbuckle, N.B., Saleem, J.J., Patterson, E.S., Flanagan, M.E., Haggstrom, D.A., Doebbeling, B.N. (2014). Sources of variation in primary care clinical workflow: Implications for the design of cognitive support. *Health Informatics Journal*, 20, 35–49 [Online]. Available at: https://api.semanticscholar.org/CorpusID:5386490.

Mittal, A., Saad, M.A., Bovik, A.C. (2016). A completely blind video integrity oracle. *IEEE Transactions on Image Processing*, 25, 289–300 [Online]. Available at: https://api.semanticscholar.org/CorpusID:11198468.

Mon, T. L.L. and Aung, S.L. (2020). GPS data enrichment for detecting driver's real-time behaviour in Myanmar. In *2020 17th International Conference on Electrical Engineering/Electronics, Computer, Telecommunications and Information Technology (ECTI-CON)*, 368–371 [Online]. Available at: https://api.semanticscholar.org/CorpusID:225835026.

Müller, W., Kuwertz, A., Mühlenberg, D., Sander, J. (2017). Semantic information fusion to enhance situational awareness in surveillance scenarios. In *2017 IEEE International Conference on Multisensor Fusion and Integration for Intelligent Systems (MFI)*, Daegu, 397–402.

Munawar, H., Ullah, F., Qayyum, S., Imran Khan, S., Mojtahedi, M. (2021). UAVs in disaster management: Application of integrated aerial imagery and convolutional neural network for flood detection. *Sustainability*, 13, 7547.

Mushtaq, M. and Mellouk, A. (2017). *Quality of Experience Paradigm in Multimedia Services: Application to OTT Video Streaming and VoIP Services*. ISTE Ltd, London and John Wiley & Sons, New York. doi: 10.1016/B978-1-78548-109-3.50008-4.

Nain, N. and Lamba, S. (2018). Oriented tracklets approach for anomalous scene detection in high density crowd. In *2018 14th International Conference on Signal-Image Technology & Internet-Based Systems (SITIS)*, Las Palmas de Gran Canaria, 435–441.

National Research Council (1992). Human factors specialists' education and utilization: Results of a survey [Online]. Available at: https://nap.nationalacademies.org/catalog/1978/human-factors-specialistseducation-and-utilization-results-of-a-survey.

Naveed, M. and Qazi, S. (2019). QoE analysis of real-time video streaming over 4G-LTE for UAV-based surveillance applications. In *INTAP* [Online]. Available at: https://api.semanticscholar.org/CorpusID:219159768.

Nesen, A. and Bhargava, B.K. (2021). Towards situational awareness with multimodal streaming data fusion: Serverless computing approach. In *Proceedings of the International Workshop on Big Data in Emergent Distributed Environments*.

New York City Police Department. (2023). NYPD announces citywide crime statistics for December 2022 [Online]. Available at: https://www.nyc.gov/site/nypd/news/p00071/nypd-citywide-crime-statistics-december-2022 [Accessed 20 September 2023].

Office of the New York State Comptroller (2022). Public safety [Online]. Available at: https://www.osc.state.ny.us/reports/finance/2022-fcr/public-safety [Accessed 25 June 2023].

Palmér, M. (2021). Pynput [Online]. Available at: https://github.com/moses-palmer/pynput.

Parashar, M. and Hariri, S. (2004). Autonomic computing: An overview. In *Unconventional Programming Paradigms* [Online]. Available at: https://api.semanticscholar.org/CorpusID:1073456.

Pavlidis, I., Morellas, V., Tsiamyrtzis, P., Harp, S.A. (2001). Urban surveillance systems: From the laboratory to the commercial world. *Proceedings of the IEEE*, 89, 1478–1497.

Perez, M., Kot, A.C., Rocha, A. (2019). Detection of real-world fights in surveillance videos. In *ICASSP 2019 – 2019 IEEE International Conference on Acoustics, Speech and Signal Processing (ICASSP)*, 2662–2666 [Online]. Available at: https://api.semanticscholar.org/CorpusID:146080078.

Popoola, O.P. and Kejun Wang, K. (2012). Video-based abnormal human behavior recognition – A review *IEEE Transactions on Systems, Man, and Cybernetics, Part C (Applications and Reviews)*, 42, 865–878 [Online]. Available at: https://api.semanticscholar.org/CorpusID:40100916.

Pramanik, A., Sarkar, S., Maiti, J. (2021). A real-time video surveillance system for traffic pre-events detection. *Accident Analysis and Prevention*, 154, 106019 [Online]. Available at: https://api.semanticscholar.org/CorpusID:232774277.

Qiao, H., Wan, X., Wan, Y., Li, S., Zhang, W. (2020). A novel change detection method for natural disaster detection and segmentation from video sequence. *Sensors (Basel, Switzerland)*, 20 [Online]. Available at: https://api.semanticscholar.org/CorpusID:221625049.

Quadros, C., Cerqueira, E., dos Santos, A.L., Lim, J., Gerla, M. (2015). Beacon-less video streaming management for vanets based on qoE and link-quality. In *2015 IFIP/IEEE International Symposium on Integrated Network Management (IM)*, 191–198 [Online]. Available at: https://api.semanticscholar.org/CorpusID:16119260.

Quadros, C., dos Santos, A.L., Gerla, M., Cerqueira, E. (2016). QoE-driven dissemination of real-time videos over vehicular networks. *Computer Communications*, 91–92, 133–147 [Online]. Available at: https://api.semanticscholar.org/CorpusID:30881629.

Quattrini, R., Pierdicca, R., Morbidoni, C. (2017). Knowledge-based data enrichment for hbim: Exploring high-quality models using the semantic-web. *Journal of Cultural Heritage*, 28, 129–139 [Online]. Available at: https://api.semanticscholar.org/CorpusID:64560800.

Ramachandra, B. and Jones, M. (2019). Street scene: A new dataset and evaluation protocol for video anomaly detection. In *2020 IEEE Winter Conference on Applications of Computer Vision (WACV)*, 2558–2567 [Online]. Available at: https://api.semanticscholar.org/CorpusID:62988676.

Rendón-Segador, F.J., Álvarez-García, J.A., Salazar-González, J.L., Tommasi, T. (2023). Crimenet: Neural structured learning using vision transformer for violence detection, In *Neural Networks: The Official Journal of the International Neural Network Society*, 161, 318–329 [Online]. Available at: https://api.semanticscholar.org/CorpusID:256564548.

Rettore, P.H.L., dos Santos, B.P., Lopes, R.R.F., Maia, G., Villas, L.A., Loureiro, A.A.F. (2020). Road data enrichment framework based on heterogeneous data fusion for its. *IEEE Transactions on Intelligent Transportation Systems*, 21, 1751–1766 [Online]. Available at: https://api.semanticscholar.org/CorpusID:214762152.

Robles-Serrano, S., Sanchez-Torres, G., Branch-Bedoya, J. (2021). Automatic detection of traffic accidents from video using deep learning techniques. *Computers*, 10(11) [Online]. Available at: https://www.mdpi.com/2073-431X/10/11/148.

Rodríguez-Silva, D.A., Adkinson-Orellana, L., González-Castaño, F.J., Armiño-Franco, I., González-Martínez, D. (2012). Video surveillance based on cloud storage. In *2012 IEEE Fifth International Conference on Cloud Computing*, 991–992 [Online]. Available at: https://api.semanticscholar.org/CorpusID:29731149.

Ruiz Jr, C.R. and San Miguel, M.J.P. (2016). Automatic flood detection using the video of static cameras [Online]. Available at: https://www.dlsu.edu.ph/wp-content/uploads/pdf/conferences/research-congress-proceedings/2016/HCT/HCT-II-05.pdf.

Salehie, M. and Tahvildari, L. (2009). Self-adaptive software: Landscape and research challenges. *ACM Transactions on Autonomous and Adaptive Systems*, 4, 14:1–14:42 [Online]. Available at: https://api.semanticscholar.org/CorpusID:207171842

Sellahewa, H. and Jassim, S. (2010). Image-quality-based adaptive face recognition. *IEEE Transactions on Instrumentation and Measurement*, 59, 805–813 [Online]. Available at: https://api.semanticscholar.org/CorpusID:18569711.

Sengar, S.S. and Mukhopadhyay, S. (2016). Moving object area detection using normalized self adaptive optical flow. *Optik*, 127, 6258–6267 [Online]. Available at: https://api.semanticscholar.org/CorpusID:124411756.

Serengil, S.I. and Ozpinar, A. (2021). Hyperextended lightface: A facial attribute analysis framework. In *2021 International Conference on Engineering and Emerging Technologies (ICEET)*. IEEE, 1–4. doi: 10.1109/ICEET53442.2021.9659697.

Shackel, B. and Richardson, S. (1991). Human factors for informatics usability [Online]. Available at: https://api.semanticscholar.org/CorpusID:60896808.

Shafahi, A., Huang, W.R., Najibi, M., Suciu, O., Studer, C., Dumitras, T., Goldstein, T. (2018). Poison frogs! targeted clean-label poisoning attacks on neural networks. In *Neural Information Processing Systems* [Online]. Available at: https://api.semanticscholar.org/CorpusID:4626477.

Shao, J., Loy, C.C., Wang, X. (2014). Scene-independent group profiling in crowd. In *2014 IEEE Conference on Computer Vision and Pattern Recognition*, Columbus, 2227–2234.

Simonyan, K. and Zisserman, A. (2014). Very deep convolutional networks for large-scale image recognition. *CoRR*, abs/1409.1556.

Singh, K., Rajora, S., Vishwakarma, D.K., Tripathi, G., Kumar, S., Walia, G.S. (2020). Crowd anomaly detection using aggregation of ensembles of fine-tuned convnets. *Neurocomputing*, 371, 188–198 [Online]. Available at: https://api.semanticscholar.org/CorpusID:202936626.

Sinno, Z. and Bovik, A.C. (2018). Large-scale study of perceptual video quality. *IEEE Transactions on Image Processing*, 28, 612–627 [Online]. Available at: https://api.semanticscholar.org/CorpusID:52285071.

Sinreich, D. (2006). An architectural blueprint for autonomic computing [Online]. Available at: https://api.semanticscholar.org/CorpusID:16909837.

Soto, C., Song, B., Roy-Chowdhury, A.K. (2009). Distributed multi-target tracking in a self-configuring camera network. In *2009 IEEE Conference on Computer Vision and Pattern Recognition*, 1486–1493 [Online]. Available at: https://api.semanticscholar.org/CorpusID:10507025.

Sultani, W., Chen, C., Shah, M. (2018). Real-world anomaly detection in surveillance videos. In *2018 IEEE/CVF Conference on Computer Vision and Pattern Recognition*, 6479–6488 [Online]. Available at: https://api.semanticscholar.org/CorpusID:1610415.

Sun, H., Liang, X., Shi, W. (2017). Vu: video usefulness and its application in large-scale video surveillance systems: An early experience. In *Proceedings of the Workshop on Smart Internet of Things* [Online]. Available at: https://api.semanticscholar.org/CorpusID:28018944.

Sutton, R.S. and Barto, A.G. (2005). Reinforcement learning: An introduction. *IEEE Transactions on Neural Networks*, 16, 285–286 [Online]. Available at: https://api.semanticscholar.org/CorpusID:9166388.

Tang, Z., Lin, Y.-S., Lee, K.-H., Hwang, J.-N., Chuang, J.-H., Fang, Z. (2016). Camera self-calibration from tracking of moving persons. In *2016 23rd International Conference on Pattern Recognition (ICPR)*, 265–270 [Online]. Available at: https://api.semanticscholar.org/CorpusID:19817607.

Tian, Y., Pang, G., Chen, Y., Singh, R., Verjans, J.W., Carneiro, G. (2021). Weakly-supervised video anomaly detection with robust temporal feature magnitude learning. In *2021 IEEE/CVF International Conference on Computer Vision (ICCV)*, 4955–4966. Available at: https://api.semanticscholar.org/CorpusID:236950792.

Tran, D., Bourdev, L.D., Fergus, R., Torresani, L., Paluri, M. (2014). Learning spatiotemporal features with 3D convolutional networks. In *2015 IEEE International Conference on Computer Vision (ICCV)*, 4489–4497 [Online]. Available at: https://api.semanticscholar.org/CorpusID:1122604.

vdoo (n.d.). Major vulnerabilities and exploit in foscam cameras [Online]. Available at: https://www.vdoo.com/blog/vdoo-has-found-major-vulnerabilities-in-foscam-cameras [Accessed 15 May 2021].

Vennam, P., Pramod, T.C., Thippeswamy, T.B., Kim, Y.-G., Pavan Kumar, B.N. (2021). Attacks and preventive measures on video surveillance systems: A review. *Applied Sciences*, 12, 5571 [Online]. Available at: https://api.semanticscholar.org/CorpusID:236235858.

Verdecia-Cabrera, A., Blanco, I.F., Díaz, A.O., Sarabia, Y.R., Diez, H.R.G. (2019). Classifier ensemble algorithm for learning from non-stationary data stream [Online]. Available at: https://api.semanticscholar.org/CorpusID:86411786.

de Walle, B.V., Brugghemans, B., Comes, T. (2016). Improving situation awareness in crisis response teams: An experimental analysis of enriched information and centralized coordination. *International Journal of Human-Computer Studies*, 95, 66–79 [Online]. Available at: https://api.semanticscholar.org/CorpusID:34760325.

Wan, S., Ding, S., Chen, C. (2021). Edge computing enabled video segmentation for real-time traffic monitoring in internet of vehicles. *Pattern Recognition*, 121, 108146 [Online]. Available at: https://api.semanticscholar.org/CorpusID:237826677.

Wang, B., Ye, M., Li, X., Zhao, F., Ding, J. (2012). Abnormal crowd behavior detection using high-frequency and spatio-temporal features. *Machine Vision and Applications – MVA*, 23.

Wang, L., Xiong, Y., Wang, Z., Qiao, Y., Lin, D., Tang, X., Gool, L.V. (2016). Temporal segment networks: Towards good practices for deep action recognition. In *European Conference on Computer Vision*, Amsterdam.

Wang, Q., Dai, H., Wu, D., Xiao, H. (2018a). Data analysis on video streaming QoE over mobile networks. In *EURASIP Journal on Wireless Communications and Networking*, 1–10 [Online]. Available at: https://api.semanticscholar.org/CorpusID:51762321.

Wang, T., Qiao, M., Deng, Y., Zhou, Y., Wang, H., Lyu, Q., Snoussi, H. (2018b). Abnormal event detection based on analysis of movement information of video sequence. *Optik*, 152, 50–60 [Online]. Available at: https://utt.hal.science/hal-03320601.

Wang, D., Yu, M., Jiang, Z., Song, Y. (2020). Blind quality assessment for high dynamic range video systems. In *Proceedings of the 2020 4th International Conference on Computer Science and Artificial Intelligence* [Online]. Available at: https://api.semanticscholar.org/CorpusID:232264255.

Williams, R.J. (1992). Simple statistical gradient-following algorithms for connectionist reinforcement learning. *Machine Learning*, 8, 229–256 [Online]. Available at: https://api.semanticscholar.org/CorpusID:2332513.

Wu, P.-H., Hwang, J.-N., Pyun, J.-Y., Lan, K.-M., Chen, J.-R. (2012). QoE-aware resource allocation for integrated surveillance system over 4G mobile networks. In *2012 IEEE International Symposium on Circuits and Systems*, 1103–1106 [Online]. Available at: https://api.semanticscholar.org/CorpusID:6402853.

Wu, P., Liu, J., Shi, Y., Sun, Y., Shao, F., Wu, Z., Yang, Z. (2020). Not only look, but also listen: Learning multimodal violence detection under weak supervision. In *European Conference on Computer Vision* [Online]. Available at: https://api.semanticscholar.org/CorpusID:220424732.

Xu, M., Yu, X., Chen, D., Wu, C., Jiang, Y. (2019). An efficient anomaly detection system for crowded scenes using variational autoencoders. *Applied Sciences*, 9(16) [Online]. Available at: https://www.mdpi.com/2076-3417/9/16/3337.

Yang, B., Cao, J., Ni, R., Zou, L. (2018a). Anomaly detection in moving crowds through spatiotemporal autoencoding and additional attention. *Advances in Multimedia*, 2018, 1–8.

Yang, F., Zhang, Q., Wang, M., Qiu, G. (2018b). Quality classified image analysis with application to face detection and recognition. In *2018 24th International Conference on Pattern Recognition (ICPR)*, 2863–2868 [Online]. Available at: https://api.semanticscholar.org/CorpusID:10300625.

Yi, C., Wu, S., Xi, B., Ming, D., Zhang, Y., Zhou, Z. (2020). Terrorist video detection system based on faster R-CNN and lightGBM. In *Proceedings of the 4th International Conference on Computer Science and Application Engineering* [Online]. Available at: https://api.semanticscholar.org/CorpusID:224804931.

Zhang, Q., Yu, Z., Shi, W., Zhong, H. (2016). Demo abstract: Evaps: Edge video analysis for public safety. In *2016 IEEE/ACM Symposium on Edge Computing (SEC)*, 121–122 [Online]. Available at: https://api.semanticscholar.org/CorpusID:16547474.

Zhang, M., Li, T., Yu, Y., Li, Y., Hui, P., Zheng, Y. (2020). Urban anomaly analytics: Description, detection, and prediction. *IEEE Transactions on Big Data*, 8, 809–826 [Online]. Available at: https://api.semanticscholar.org/CorpusID:216553715.

Zhao, L., Li, Q., Wang, P., Wang, J., Wu, E. (2020). Activedeeper: A model-based active data enrichment system. In *Proceedings of the VLDB Endowment*, 13, 2885–2888 [Online]. Available at: https://api.semanticscholar.org/CorpusID:221539360.

Zhou, Z., Dong, X., Li, Z., Yu, K., Ding, C., Yang, Y. (2022). Spatio-temporal feature encoding for traffic accident detection in vanet environment. *IEEE Transactions on Intelligent Transportation Systems*, 23(10), 19772–19781.

Zivkovic, Z. (2004). Improved adaptive Gaussian mixture model for background subtraction. In *Proceedings of the 17th International Conference on Pattern Recognition, 2004. ICPR 2004*, Cambridge, 28–31.

Zivkovic, Z. and van der Heijden, F. (2006). Efficient adaptive density estimation per image pixel for the task of background subtraction. *Pattern Recognition Letters*, 27, 773–780.

Index

Q

R, S

T

U, V

Other titles from

iSTE

in

Networks and Telecommunications

2024

HORVÁTH András, TELEK Miklós
Phase Type Distributions: Theory and Application
(Stochastic Models in Computer Science and Telecommunication Networks Set – Volume 2)

2023

RAVALIMINOARIMALALASON Toky Basilide, RANDIMBINDRAINIBE Falimanana
Queues Applied to Telecoms: Courses and Exercises

TONG Van Van, SOUIHI Sami, TRAN Hai-Anh, MELLOUK Abdelhamid
Troubleshooting for Network Operators: The Road to a New Paradigm with Encrypted Traffic
(New Generation Networks Set – Volume 3)

2022

BANNOUR Fetia, SOUIHI Sami, MELLOUK Abdelhamid
Software-Defined Networking: Extending SDN Control to Large-Scale Networks
(New Generation Networks Set – Volume 2)

BENAROUS Leila, BITAM Salim, MELLOUK Abdelhamid
Security in Vehicular Networks: Focus on Location and Identity Privacy
(New Generation Networks Set – Volume 1)

2021

LAUNAY Frédéric
NG-RAN and 5G-NR: 5G Radio Access Network and Radio Interference

2020

GONTRAND Christophe
Digital Communication Techniques

PUJOLLE Guy
Software Networks: Virtualization, SDN, 5G and Security (2nd edition revised and updated)
(Advanced Network Set – Volume 1)

2019

LAUNEY Frédéric, PEREZ André
LTE Advanced Pro: Towards the 5G Mobile Network
Harmonic Concept and Applications

TOUNSI Wiem
Cyber-Vigilance and Digital Trust: Cyber Security in the Era of Cloud Computing and IoT

2018

ANDÍA Gianfranco, DURO Yvan, TEDJINI Smail
Non-linearities in Passive RFID Systems: Third Harmonic Concept and Applications

BOUILLARD Anne, BOYER Marc, LE CORRONC Euriell
Deterministic Network Calculus: From Theory to Practical Implementation

LAUNAY Frédéric, PEREZ André
LTE Advanced Pro: Towards the 5G Mobile Network

PEREZ André
Wi-Fi Integration to the 4G Mobile Network

2017

BENSLAMA Malek, BENSLAMA Achour, ARIS Skander
Quantum Communications in New Telecommunications Systems

HILT Benoit, BERBINEAU Marion, VINEL Alexey, PIROVANO Alain
Networking Simulation for Intelligent Transportation Systems: High Mobile Wireless Nodes

LESAS Anne-Marie, MIRANDA Serge
The Art and Science of NFC Programming
(Intellectual Technologies Set – Volume 3)

2016

AL AGHA Khaldoun, PUJOLLE Guy, ALI-YAHIYA Tara
Mobile and Wireless Networks
(Advanced Network Set – Volume 2)

BATTU Daniel
Communication Networks Economy

BENSLAMA Malek, BATATIA Hadj, MESSAI Abderraouf
Transitions from Digital Communications to Quantum Communications: Concepts and Prospects

CHIASSERINI Carla Fabiana, GRIBAUDO Marco, MANINI Daniele
Analytical Modeling of Wireless Communication Systems (Stochastic Models in Computer Science and Telecommunication Networks Set – Volume 1)

EL FALLAH SEGHROUCHNI Amal, ISHIKAWA Fuyuki, HÉRAULT Laurent, TOKUDA Hideyuki
Enablers for Smart Cities

PEREZ André
VoLTE and ViLTE

2015

BENSLAMA Malek, BATATIA Hadj, BOUCENNA Mohamed Lamine
Ad Hoc Networks Telecommunications and Game Theory

BENSLAMA Malek, KIAMOUCHE Wassila, BATATIA Hadj
Connections Management Strategies in Satellite Cellular Networks

BERTHOU Pascal, BAUDOIN Cédric, GAYRAUD Thierry, GINESTE Matthieu
Satellite and Terrestrial Hybrid Networks

CUADRA-SANCHEZ Antonio, ARACIL Javier
Traffic Anomaly Detection

LE RUYET Didier, PISCHELLA Mylène
Digital Communications 1: Source and Channel Coding

PEREZ André
LTE and LTE Advanced: 4G Network Radio Interface

PISCHELLA Mylène, LE RUYET Didier
Digital Communications 2: Digital Modulations

2014

ANJUM Bushra, PERROS Harry
Bandwidth Allocation for Video under Quality of Service Constraints

BATTU Daniel
New Telecom Networks: Enterprises and Security

BEN MAHMOUD Mohamed Slim, GUERBER Christophe, LARRIEU Nicolas, PIROVANO Alain, RADZIK José
Aeronautical Air−Ground Data Link Communications

BITAM Salim, MELLOUK Abdelhamid
Bio-inspired Routing Protocols for Vehicular Ad-Hoc Networks

CAMPISTA Miguel Elias Mitre, RUBINSTEIN Marcelo Gonçalves
Advanced Routing Protocols for Wireless Networks

CHETTO Maryline
Real-time Systems Scheduling 1: Fundamentals
Real-time Systems Scheduling 2: Focuses

EXPOSITO Ernesto, DIOP Codé
Smart SOA Platforms in Cloud Computing Architectures

LARRIEU Nicolas, VARET Antoine
Rapid Prototyping of Software for Avionics Systems: Model-oriented Approaches for Complex Systems Certification

MELLOUK Abdelhamid, CUADRA-SANCHEZ Antonio
Quality of Experience Engineering for Customer Added Value Services

OTEAFY Sharief M.A., HASSANEIN Hossam S.
Dynamic Wireless Sensor Networks

PEREZ André
Network Security

PERRET Etienne
Radio Frequency Identification and Sensors: From RFID to Chipless RFID

REMY Jean-Gabriel, LETAMENDIA Charlotte
LTE Standards
LTE Services

RIGO Michel
Formal Languages, Automata and Numeration Systems 1: Introduction to Combinatorics on Words
Formal Languages, Automata and Numeration Systems 2: Applications to Recognizability and Decidability

TANWIR Savera, PERROS Harry
VBR Video Traffic Models

VAN METER Rodney
Quantum Networking

XIONG Kaiqi
Resource Optimization and Security for Cloud Services

2013

ASSING Dominique, CALÉ Stéphane
Mobile Access Safety: Beyond BYOD

BEN MAHMOUD Mohamed Slim, LARRIEU Nicolas, PIROVANO Alain
Risk Propagation Assessment for Network Security: Application to Airport Communication Network Design

BERTIN Emmanuel, CRESPI Noël
Architecture and Governance for Communication Services

BEYLOT André-Luc, LABIOD Houda
Vehicular Networks: Models and Algorithms

BRITO Gabriel M., VELLOSO Pedro Braconnot, MORAES Igor M.
Information-Centric Networks: A New Paradigm for the Internet

DEUFF Dominique, COSQUER Mathilde
User-Centered Agile Method

DUARTE Otto Carlos, PUJOLLE Guy
Virtual Networks: Pluralistic Approach for the Next Generation of Internet

FOWLER Scott A., MELLOUK Abdelhamid, YAMADA Naomi
LTE-Advanced DRX Mechanism for Power Saving

FERRANT Jean-Loup *et al.*
Synchronous Ethernet and IEEE 1588 in Telecoms: Next Generation Synchronization Networks

MELLOUK Abdelhamid, HOCEINI Said, TRAN Hai Anh
Quality-of-Experience for Multimedia: Application to Content Delivery Network Architecture

NAIT-SIDI-MOH Ahmed, BAKHOUYA Mohamed, GABER Jaafar, WACK Maxime
Geopositioning and Mobility

PEREZ André
Voice over LTE: EPS and IMS Networks

TERRE Michel, PISCHELLA Mylène, VIVIER Emmanuelle
Wireless Telecommunication Systems

2012

AL AGHA Khaldoun
Network Coding

BOUCHET Olivier
Wireless Optical Communications

DECREUSEFOND Laurent, MOYAL Pascal
Stochastic Modeling and Analysis of Telecoms Networks

DUFOUR Jean-Yves
Intelligent Video Surveillance Systems

EXPOSITO Ernesto
Advanced Transport Protocols: Designing the Next Generation

FOULADGAR Javad
Electrothermics

JUMIRA Oswald, ZEADALLY Sherali
Energy Efficiency in Wireless Networks

KRIEF Francine
Green Networking

PEREZ André
Mobile Networks Architecture

POLER Raúl, CARNEIRO Luis Maia, JASINSKI Thomas, ZOLGHADRI Marc, PEDRAZZOLI Paolo
Intelligent Non-hierarchical Manufacturing Networks

2011

BONALD Thomas, FEUILLET Mathieu
Network Performance Analysis

CARBOU Romain, DIAZ Michel, EXPOSITO Ernesto, ROMAN Rodrigo
Digital Home Networking

CHABANNE Hervé, URIEN Pascal, SUSINI Jean-Ferdinand
RFID and the Internet of Things

GARDUNO BARRERA David, DIAZ Michel
Communicating Systems with UML 2: Modeling and Analysis of Network Protocols

LAHEURTE Jean-Marc
Compact Antennas for Wireless Communications and Terminals: Theory and Design

PALICOT Jacques
Radio Engineering: From Software Radio to Cognitive Radio

PEREZ André
IP, Ethernet and MPLS Networks: Resource and Fault Management

RÉMY Jean-Gabriel, LETAMENDIA Charlotte
Home Area Networks and IPTV

TOUTAIN Laurent, MINABURO Ana
Local Networks and the Internet: From Protocols to Interconnection

2010

BEGAUD Xavier
Ultra Wide Band Antennas

CHAOUCHI Hakima
The Internet of Things

FRIKHA Mounir
Ad Hoc Networks: Routing, QoS and Optimization

KRIEF Francine
Communicating Embedded Systems / Network Applications

2009

CHAOUCHI Hakima, MAKNAVICIUS Maryline
Wireless and Mobile Network Security

VIVIER Emmanuelle
Radio Resources Management in WiMAX

2008

CHADUC Jean-Marc, POGOREL Gérard
The Radio Spectrum

GAÏTI Dominique
Autonomic Networks

LABIOD Houda
Wireless Ad Hoc and Sensor Networks

LECOY Pierre
Fiber-optic Communications

MELLOUK Abdelhamid
End-to-End Quality of Service Engineering in Next Generation Heterogeneous Networks

PAGANI Pascal *et al.*
Ultra-wideband Radio Propagation Channel

2007

BENSLIMANE Abderrahim
Multimedia Multicast on the Internet

PUJOLLE Guy
Management, Control and Evolution of IP Networks

SANCHEZ Javier, THIOUNE Mamadou
UMTS

VIVIER Guillaume
Reconfigurable Mobile Radio System

Printed and bound by CPI Group (UK) Ltd, Croydon, CR0 4YY

13/09/2026

14961013-0003